PLUMBING
PROFESSIONAL REFERENCE

Paul Rosenberg

Created exclusively
for DeWALT by:

www.dewalt.com/guides

OTHER TITLES AVAILABLE

Trade Reference Series

- Blueprint Reading
- Construction
- Construction Estimating
- Construction Safety/OSHA
- Datacom
- Electric Motor
- Electrical
- Electrical Estimating
- HVAC/R – Master Edition
- Lighting & Maintenance
- Residential Remodeling & Repair
- Security, Sound & Video
- Spanish/English Construction Dictionary – Illustrated
- Wiring Diagrams

Exam and Certification Series

- Building Contractor's Licensing Exam Guide
- Electrical Licensing Exam Guide
- HVAC Technician Certification Exam Guide
- Plumbing Licensing Exam Guide

For a complete list of The DeWALT Professional Trade Reference Series visit **www.dewalt.com/guides**.

This Book Belongs To:

Name:_____

Company: _____

Title: _____

Department: _____

Company Address: _____

_____

_____

Company Phone: _____

Home Phone: _____

Pal Publications, Inc.
800 Heritage Drive, Suite 810
Pottstown, PA 19464-3810

Copyright © 2005 by Pal Publications, Inc.
First edition published 2005

NOTICE OF RIGHTS
All rights reserved. No part of this book may be reproduced or transmitted in any form or by any means, electronic or mechanical including photocopying, recording or by any information storage and retrieval system, without permission in writing from the publisher.

NOTICE OF LIABILITY
Information contained in this work has been obtained from sources believed to be reliable. However, neither Pal Publications, Inc. nor its authors guarantee the accuracy or completeness of any information published herein, and neither Pal Publications, Inc. nor its authors shall be responsible for any errors, omissions, damages, liabilities or personal injuries arising out of use of this information. This work is published with the understanding that Pal Publications, Inc. and its authors are supplying information but are not attempting to render engineering or other professional services. If such services are required, the assistance of an appropriate professional should be sought. The reader is expressly warned to consider and adopt all safety precautions and to avoid all potential hazards. The publisher makes no representation or warranties of any kind, nor are any such representations implied with respect to the material set forth here. The publisher shall not be liable for any special, consequential, or exemplary damages resulting, in whole or part, from the readers' use of, or reliance upon, this material.

ISBN-13: 978-0-9770003-1-9
ISBN-10: 0-9770003-1-1

09 08 07 06 05   5 4 3 2 1
Printed in Canada

DEWALT is a registered trademark of DEWALT Industrial Tool Co., used under license. All rights reserved. The yellow and black color scheme is a trademark for DEWALT Power Tools and Accessories. Pal Publications, Inc., a licensee of DEWALT Industrial Tools. 800 Heritage Drive, Suite 810, Pottstown, PA 19464, Tel.: 800-246-2175.

A Note To Our Customers

We have manufactured this book to the highest quality standards possible. The cover is made of a flexible, durable and water-resistant material able to withstand the toughest on-the-job conditions. We also utilize a special binding process which allows this book to lay flatter than traditional paperback books that tend to snap shut while in use.

Preface

Plumbing is one of the largest and certainly one of the most essential of the construction trades. Yet it gets much less coverage than most of the others. So, putting together this book was a pleasure for me. It was also gratifying to see that the book developed as we had hoped — a small book with almost everything a plumber really needs to know for his or her daily work.

Obviously, any attempt to cover the entire plumbing and piping industry in a single book requires that certain material is excluded. In this book I chose to cover the material that is used most by people who install and maintain plumbing systems. Bear in mind that unlike many other construction specialties, there are a number of plumbing codes, and their acceptance varies from place to place. In the design data, we used the most stringent of the requirements where we found differences. Generally, the various codes run parallel to one another, but we defaulted to the most demanding one in any particular case. You may wish to verify your local codes especially in cases where you are installing a large number of items. The cost difference between a single 2" or 2½" fitting may be negligible, but the difference between a thousand 2" or 2½" fittings is significant.

Naturally, there may be some aspects of plumbing work that I have overlooked, or that are not covered in sufficient depth for some readers. I monitor the industry and will update this book on a regular basis. I will also attempt to include additional material suggested by readers and to keep pace with developments in the trade.

Best wishes,
Paul Rosenberg

Chapter 1 –
Pipes, Fittings and Valves 1-1

Chapter 2 – *Pipefitting* 2-1

Chapter 3 – *Design Data* 3-1

Chapter 4 – *Drainage Systems* 4-1

Chapter 9 – *Conversion Factors* 9-1

Chapter 10 –
Symbols and Abbreviations **10-1**

Chapter 11 – *Glossary*. **11-1**

CHAPTER 1
Pipe, Fittings and Valves

In plumbing, the pipe size measurement given is always referred to as nominal pipe size (N.P.S.) and is the measurement of the inside diameter (ID). In the air conditioning and refrigeration trades, pipe and tubing are referenced by their outside diameter (OD) measurement.

For example, a ¾ inch copper pipe in plumbing would be called a ⅞ inch copper pipe in the HVACR trade.

COMMON PLUMBING MEASUREMENT TERMS

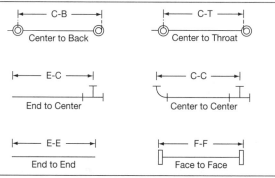

HEAVY FLUID SUPPORT MULTIPLIERS

For heavier fluids, multiply the support spacing on pages 1-5 and 1-8 by the following correction factors:

Specific Gravity of Fluid	1.00	1.10	1.20	1.40	1.60	2.00	2.50
Correction Factor	1.00	.98	.96	.93	.90	.85	.80

For insulated lines, reduce the spans by 70% of the given values.

PLASTIC PIPE

PVC and ABS are the most common types of plastic pipe. However, the correct primers and solvents must be used on each type or the joints will not seal properly and overall strength will be diminished.

Type	Characteristics and Guidelines
PVC	**Polyvinyl Chloride, Type 1, Grade 1**. Strong, rigid and resistant to a variety of caustic fluids. PVC is easy to work with and readily available. Maximum useable temperature is 140°F and pressure ratings start at a minimum of 125 to 200 psi. PVC can be used with water, gas, and drainage systems. DO NOT use with hot water systems.
ABS	**Acrylonitrile Butadiene Styrene, Type 1**. Like PVC, ABS is strong, rigid and resistant to a variety of caustic fluids and is very common, easy to work with and readily available. Maximum useable temperature is 160°F at low pressure. ABS is generally used as a DWV (drain, waste and vent) line.
CPVC	**Chlorinated polyvinyl chloride**. Similar to PVC but designed specifically for piping water at up to 180°F. CPVC has a pressure rating of 100 psi.
PE and **PEX**	**Polyethylene and Polyethylene cross-linked**. Both are flexible pipes normally used for low pressure water systems such as sprinklers. DO NOT use PE (polyethylene) pipe with hot water systems.
PB	**Polybutylene**. A flexible pipe for pressurized water systems both hot and cold. Use only compression and banded type joints for installation.
Polypropylene	A lightweight material that is rated to 180°F at low pressure. Highly resistant to caustic fluids such as acids, bases, and solvents. Recommended for laboratory plumbing applications.
PVDF	**Polyvinylidene fluoride**. Strong, durable, and resistant to abrasion, acids, bases and solvents that is rated to 280°F. Recommended for laboratory plumbing applications.
FRP Epoxy	A thermosetting plastic over fiberglass. Very strong with excellent caustic fluid resistance properties and is rated to 220°F. Also used in laboratory applications.

PLASTIC PIPE DIMENSIONS AND WEIGHTS

SCHEDULE 40

Nominal Pipe Size (in.)	Outside Diameter (in.)	PVC		CPVC	
		Wall Thickness (in.)	Weight lbs. per ft.	Wall Thickness (in.)	Weight lbs. per ft.
¼	0.540	–	–	–	–
½	0.840	0.109	0.16	0.109	0.19
¾	1.050	0.113	0.22	0.113	0.25
1	1.315	0.133	0.32	0.133	0.38
1¼	1.660	0.140	0.43	0.140	0.51
1½	1.900	0.145	0.52	0.145	0.61
2	2.375	0.154	0.70	0.154	0.82
2½	2.875	0.203	1.10	0.203	1.29
3	3.500	0.216	1.44	0.216	1.69
4	4.500	0.237	2.05	0.237	2.33
6	6.625	0.280	3.61	0.280	4.10
8	8.625	0.322	5.45	–	–
10	10.75	0.365	7.91	–	–
12	12.75	0.406	10.35	–	–

PLASTIC PIPE DIMENSIONS AND WEIGHTS (*cont.*)

SCHEDULE 80

Nominal Pipe Size (in.)	Outside Diameter (in.)	PVC Wall Thickness (in.)	PVC Weight lbs. per ft.	CPVC Wall Thickness (in.)	CPVC Weight lbs. per ft.
¼	0.540	0.119	0.10	0.119	0.12
½	0.840	0.147	0.21	0.147	0.24
¾	1.050	0.154	0.28	0.154	0.33
1	1.315	0.179	0.40	0.179	0.49
1¼	1.660	0.191	0.57	0.191	0.67
1½	1.900	0.200	0.69	0.200	0.81
2	2.375	0.218	0.95	0.218	1.09
2½	2.875	0.276	1.45	0.276	1.65
3	3.500	0.300	1.94	0.300	2.21
4	4.500	0.337	2.83	0.337	3.23
6	6.625	0.432	5.41	0.432	6.17
8	8.625	0.500	8.22	0.500	9.06
10	10.750	0.593	12.28	–	–
12	12.750	0.687	17.10	–	–

Nominal Pipe Size (in.)	Outside Diameter (in.)	PVDF Wall Thickness (in.)	PVDF Weight lbs. per ft.	Polypropylene Wall Thickness (in.)	Polypropylene Weight lbs. per ft.
½	0.840	0.147	0.24	0.147	0.14
¾	1.050	0.154	0.33	0.154	0.19
1	1.315	0.179	0.49	0.179	0.27
1¼	1.660	0.191	–	0.191	0.38
1½	1.900	0.200	0.81	0.200	0.45
2	2.375	0.218	1.13	0.218	0.62

PVC PIPE SUPPORT SPACING IN FEET

SCHEDULE 40

Nominal Pipe Size (in.)	Temperature Range (F)				
	60°	80°	100°	120°	140°
1/4	3.75	3.50	3.00	2.50	2.00
1/2	4.25	4.00	3.50	3.00	2.50
3/4	4.50	4.25	4.00	3.50	3.00
1	5.00	4.75	4.50	3.75	3.25
1 1/4	5.25	5.00	4.75	4.00	3.50
1 1/2	5.50	5.25	5.00	4.25	3.75
2	6.00	5.50	5.00	4.50	4.00
2 1/2	6.75	6.25	5.75	4.75	4.25
3	7.25	6.75	6.25	5.25	4.50
4	7.75	7.50	6.75	6.00	4.75
6	8.75	8.50	7.75	6.50	5.25
8	9.75	9.25	8.50	7.75	6.00
10	10.25	9.75	9.00	8.00	6.75
12	11.00	10.25	9.75	8.25	7.25

SCHEDULE 80

Nominal Pipe Size (in.)	Temperature Range (F)				
	60°	80°	100°	120°	140°
1/4	4.25	4.00	3.75	3.00	2.75
1/2	4.50	4.25	4.00	3.75	3.00
3/4	4.75	4.50	4.25	4.00	3.50
1	5.00	4.75	4.50	4.25	3.75
1 1/4	6.00	5.00	4.75	4.50	4.25
1 1/2	6.50	5.50	5.25	5.00	4.75
2	6.75	5.75	5.50	5.25	5.00
2 1/2	7.25	6.75	5.75	5.50	5.25
3	7.75	7.50	7.00	6.25	5.50
4	9.00	8.75	7.25	6.50	5.75
6	9.75	9.50	8.50	7.75	6.50
8	11.00	10.25	9.75	8.75	7.00
10	11.50	10.50	10.25	9.50	7.75
12	12.50	12.25	11.50	10.25	8.75

STEEL PIPE DIMENSIONS AND WEIGHTS

SCHEDULE 40

Nominal Pipe Size (in.)	Outside Diameter (in.)	Wall Thickness (in.)	Inside Diameter (in.)	Pipe Weight lbs. per ft.
1/8	.405	.068	.269	.245
1/4	.540	.088	.364	.425
3/8	.675	.091	.493	.568
1/2	.840	.109	.622	.851
3/4	1.050	.113	.824	1.131
1	1.315	.133	1.049	1.679
1 1/4	1.660	.140	1.380	2.273
1 1/2	1.900	.145	1.610	2.718
2	2.375	.154	2.067	3.653
2 1/2	2.875	.203	2.469	5.793
3	3.500	.216	3.068	7.580
3 1/2	4.000	.226	3.548	9.110
4	4.500	.237	4.026	10.790
5	5.563	.258	5.047	14.620
6	6.625	.280	6.065	18.970
8	8.625	.322	7.981	28.550
10	10.750	.365	10.020	40.480
12	12.750	.375	12.000	49.560
14	14.000	.500	13.000	72.090
16	16.000	.500	15.000	82.770
18	18.000	.562	16.876	104.750
20	20.000	.593	18.814	122.910

STEEL PIPE DIMENSIONS AND WEIGHTS (cont.)

SCHEDULE 80

Nominal Pipe Size (in.)	Outside Diameter (in.)	Wall Thickness (in.)	Inside Diameter (in.)	Pipe Weight lbs. per ft.
$\frac{1}{8}$	.405	.095	.215	.315
$\frac{1}{4}$	.540	.119	.302	.535
$\frac{3}{8}$	.675	.126	.423	.739
$\frac{1}{2}$	.840	.147	.546	1.088
$\frac{3}{4}$	1.050	.154	.742	1.474
1	1.315	.179	.957	2.172
$1\frac{1}{4}$	1.660	.191	1.278	2.997
$1\frac{1}{2}$	1.900	.200	1.500	3.631
2	2.375	.218	1.939	5.022
$2\frac{1}{2}$	2.875	.276	2.323	7.661
3	3.500	.300	2.900	10.250
$3\frac{1}{2}$	4.000	.318	3.364	12.510
4	4.500	.337	3.826	14.980
5	5.563	.375	4.813	20.780
6	6.625	.432	5.761	28.570
8	8.625	.500	7.625	43.390
10	10.750	.500	9.750	54.740
12	12.750	.500	11.750	65.420
14	14.000	.750	12.500	106.130
16	16.000	.843	14.314	136.460
18	18.000	.937	16.126	170.750
20	20.000	1.031	17.938	208.870

PIPE THREADING DIMENSIONS (NPT)

Nominal Pipe Size (in.)	Threads Per Inch	Approximate Length of Thread (in.)	Approximate Number of Threads To Be Cut	Approximate Total Thread Makeup, Hand and Wrench (in.)
$\frac{1}{8}$	27	$\frac{3}{8}$	10	$\frac{1}{4}$
$\frac{1}{4}$	18	$\frac{5}{8}$	11	$\frac{3}{8}$
$\frac{3}{8}$	18	$\frac{5}{8}$	11	$\frac{3}{8}$
$\frac{1}{2}$	14	$\frac{3}{4}$	10	$\frac{7}{16}$
$\frac{3}{4}$	14	$\frac{3}{4}$	10	$\frac{1}{2}$
1	$11\frac{1}{2}$	$\frac{7}{8}$	10	$\frac{9}{16}$
$1\frac{1}{4}$	$11\frac{1}{2}$	1	11	$\frac{9}{16}$
$1\frac{1}{2}$	$11\frac{1}{2}$	1	11	$\frac{9}{16}$
2	$11\frac{1}{2}$	1	11	$\frac{5}{8}$
$2\frac{1}{2}$	8	$1\frac{1}{2}$	12	$\frac{7}{8}$
3	8	$1\frac{1}{2}$	12	1
$3\frac{1}{2}$	8	$1\frac{5}{8}$	13	$1\frac{1}{16}$
4	8	$1\frac{5}{8}$	13	$1\frac{1}{16}$
5	8	$1\frac{3}{4}$	14	$1\frac{3}{16}$
6	8	$1\frac{3}{4}$	14	$1\frac{3}{16}$
8	8	$1\frac{7}{8}$	15	$1\frac{5}{16}$
10	8	2	16	$1\frac{1}{2}$
12	8	$2\frac{1}{8}$	17	$1\frac{5}{8}$

STEEL PIPE SUPPORT SPACING IN FEET

Nominal Pipe Size (in.)	Support Spacing (ft.)	Nominal Pipe Size (in.)	Support Spacing (ft.)
1	7	4	14
$1\frac{1}{2}$	9	6	17
2	10	8	19
3	12	10	22

COPPER PLUMBING PIPE AND TUBING

When installing copper pipe, sweat fittings are measured by their inside diameter (ID) and compression fittings are measured by their outside diameter (OD). Always use a 50/50 solid core solder along with a high quality flux when soldering sweat fittings. DO NOT use a rosin core type solder.

Type	Characteristics and Guidelines
DWV	**Drain, Waste and Vent** is recommended for above ground use only. Not to be used in pressure applications and install only using sweat fittings. Available in hard type from 1¼ inch to 6 inch sizes.
K	Flexible copper tubing with a thicker wall than Type L and M. Required for all underground installations. Uses include plumbing, heating, steam, gas and oil where thick walled tubing is required. Can be used with sweat, flared and compression fittings. Available in hard and soft types.
L	Standard copper tubing used for interior, above ground applications including air conditioning, heating, steam, gas and oil. Because of its flexibility, be very careful not to crimp the line when bending. Tools are available to make bending safer and easier. Sweat, compression and flare fittings are available. DO NOT use compression fittings for gas lines. Available in hard and soft types.
M	Generally used with interior heating and pressure line applications. Wall thickness is less than types K and L. Install with sweat fittings. Available in hard and soft types.

COPPER PLUMBING PIPE AND TUBING DIMENSIONS AND WEIGHTS

Nominal Pipe Size (in.)	Outside Diameter (in.) All Types	Wall Thickness (in.)				Inside Diameter (in.)				Pounds Per Foot			
		DWV	K	L	M	DWV	K	L	M	DWV	K	L	M
¼	.375	–	.035	.030	.025	–	.305	.315	.325	–	.145	.126	.106
⅜	.500	–	.049	.035	.025	–	.402	.430	.450	–	.269	.198	.145
½	.625	–	.049	.040	.028	–	.527	.545	.569	–	.344	.285	.204
⅝	.750	–	.049	.042	.030	–	.652	.666	.690	–	.418	.362	.263
¾	.875	–	.065	.045	.032	–	.745	.785	.811	–	.641	.455	.328
1	1.125	–	.065	.050	.035	–	.995	1.025	1.055	–	.839	.655	.465
1¼	1.375	.040	.065	.055	.042	1.295	1.245	1.265	1.291	.650	1.040	.884	.682
1½	1.625	.042	.072	.060	.049	1.541	1.481	1.505	1.527	.809	1.360	1.140	.940
2	2.125	.042	.083	.070	.058	2.041	1.959	1.985	2.009	1.070	2.060	1.750	1.460
2½	2.625	–	.095	.080	.065	–	2.435	2.465	2.495	–	2.930	2.480	2.030
3	3.125	.045	.109	.090	.072	3.035	2.907	2.945	2.981	1.690	4.000	3.330	2.680
3½	3.625	–	.120	.100	.083	–	3.385	3.425	3.459	–	5.120	4.290	3.580
4	4.125	.058	.134	.110	.095	4.009	3.857	3.905	3.935	2.870	6.510	5.380	4.660
5	5.125	.072	.160	.125	.109	4.981	4.805	4.875	4.907	4.430	9.670	7.610	6.660
6	6.125	.083	.192	.140	.122	5.959	5.741	5.845	5.881	6.100	13.900	10.200	8.920
8	8.125	.109	.271	.200	.170	7.907	7.583	7.725	7.785	10.600	25.900	19.300	16.500
10	10.125	–	.338	.250	.212	–	9.449	9.625	9.701	–	40.300	30.100	25.600
12	12.125	–	.405	.280	.254	–	11.315	11.565	11.617	–	57.800	40.400	36.700

COPPER PLUMBING PIPE AND TUBING LENGTHS

Type	Drawn (hard)	Annealed (soft)
DWV	**Straight Lengths:** All diameters 20 ft.	N/A

K	**Straight Lengths:**
	Up to 8-inch diameter 20 ft.
	10-inch diameter 18 ft.
	12-inch diameter 12 ft.
		Straight Lengths:
		Up to 8-inch diameter 20 ft.
		10-inch diameter 18 ft.
		12-inch diameter 12 ft.
		
		Coils:
		Up to 1-inch diameter 60 ft.
		  100 ft.
		1¼ and 1½-inch diameter 60 ft.
		  40 ft.
		2-inch diameter 45 ft.

L	**Straight Lengths:**
	Up to 10-inch diameter 20 ft.
	12-inch diameter 18 ft.
		Straight Lengths:
		Up to 10-inch diameter 20 ft.
		12-inch diameter 18 ft.
		
		Coils:
		Up to 1-inch diameter 60 ft.
		  100 ft.
		1¼ and 1½-inch diameter 60 ft.
		  40 ft.
		2-inch diameter 45 ft.

M	**Straight Lengths:** All diameters 20 ft.	N/A

ROLL GROOVE SPECIFICATIONS
FOR COPPER TUBING

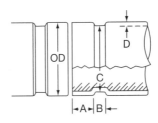

Nominal Copper Tube Size (in.)	OD	A	B	C	D
	Tube Distance Diameter (in.)	Gasket Seat (in.)	Groove Width (in.)	Groove Diameter (in.)	Groove Depth (in.)
2	2.125	.610	.300	2.029	.048
2½	2.625	.610	.300	2.525	.050
3	3.125	.610	.300	3.025	.050
4	4.125	.610	.300	4.019	.053
5	5.125	.610	.300	5.019	.053
6	6.125	.610	.300	5.999	.063

STANDARD CUT GROOVE SPECIFICATIONS

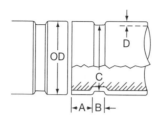

	OD	A	B	C	D
Nominal Pipe Size (in.)	Pipe Outside Diameter (in.)	Gasket Seal (in.)	Groove Width (in.)	Groove Diameter (in.)	Groove Depth (in.)
¾	1.050	.625	.313	.938	.056
1	1.315	.625	.313	1.189	.063
1¼	1.660	.625	.313	1.534	.063
1½	1.900	.625	.313	1.774	.063
2	2.375	.625	.313	2.249	.063
2½	2.875	.625	.313	2.719	.078
3	3.500	.625	.313	3.344	.078
4	4.500	.625	.375	4.344	.083
5	5.563	.625	.375	5.395	.084
6	6.625	.625	.375	6.455	.085
8	8.625	.750	.438	8.441	.092
10	10.750	.750	.500	10.562	.094
12	12.750	.750	.500	12.532	.109

LINEAR EXPANSION OF PIPING

Temp. (F)	Inches of Expansion per 100 Feet of Pipe			
	Copper	Steel	Cast Iron	Wrought Iron
-30°	–	–	–	–
-20°	.105	.072	.062	.073
-10°	.211	.145	.124	.147
0°	.316	.215	.186	.221
10°	.428	.291	.251	.298
20°	.541	.367	.317	.376
30°	.654	.442	.383	.454
40°	.767	.517	.449	.533
50°	.880	.592	.515	.612
60°	.993	.667	.581	.691
70°	1.107	.742	.647	.770
80°	1.221	.817	.713	.849
90°	1.335	.892	.779	.928
100°	1.449	.968	.845	1.007
110°	1.565	1.048	.915	1.090
120°	1.681	1.128	.985	1.174
130°	1.797	1.208	1.056	1.258
140°	1.913	1.287	1.127	1.342
150°	2.029	1.366	1.198	1.426
160°	2.145	1.445	1.269	1.510
170°	2.261	1.524	1.340	1.594
180°	2.377	1.603	1.411	1.678

LINEAR EXPANSION OF PIPING (cont.)

Temp. (F)	Inches of Expansion per 100 Feet of Pipe			
	Copper	Steel	Cast Iron	Wrought Iron
190°	2.494	1.682	1.482	1.762
200°	2.611	1.761	1.553	1.846
210°	2.727	1.843	1.626	1.931
220°	2.843	1.925	1.699	2.016
230°	2.959	2.008	1.773	2.101
240°	3.075	2.091	1.847	2.186
250°	3.191	2.174	1.921	2.271
260°	3.308	2.257	1.995	2.356
270°	3.425	2.340	2.069	2.441
280°	3.542	2.423	2.143	2.526
290°	3.659	2.506	2.217	2.612
300°	3.776	2.589	2.291	2.698
310°	3.896	2.674	2.368	2.787
320°	4.016	2.759	2.445	2.876
330°	4.136	2.844	2.522	2.965
340°	4.256	2.929	2.599	3.054
350°	4.376	3.015	2.676	3.143
360°	4.497	3.101	2.754	3.232
370°	4.618	3.187	2.832	3.321
380°	4.739	3.273	2.910	3.410
390°	4.860	3.359	2.988	3.499
400°	4.981	3.445	3.066	3.589

FITTING DIAGRAMS

1/16 Bend

1/8 Bend

1/6 Bend

1/4 Bend

Long 1/8 Bend

Long Sweep

Sanitary Tee

Y

Combination Tee

FITTING DIAGRAMS (cont.)

Tap Short
1/4 Bend

Tap Sanitary Tee

Tap Extension Piece

Tap Adapter

Closet Flange
(Slot & Notch)

Hub Adapter

P Trap

Short Reducer

Test Tee Less Plug

Blind Plug

FITTING DIAGRAMS (cont.)

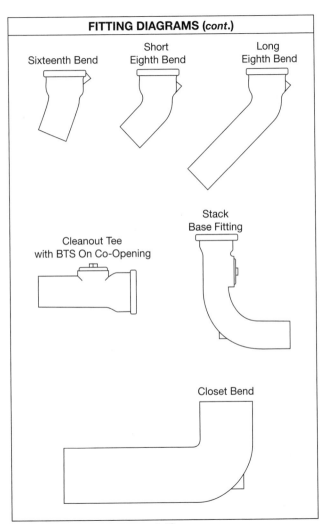

Sixteenth Bend

Short Eighth Bend

Long Eighth Bend

Cleanout Tee with BTS On Co-Opening

Stack Base Fitting

Closet Bend

FITTING DIAGRAMS (cont.)

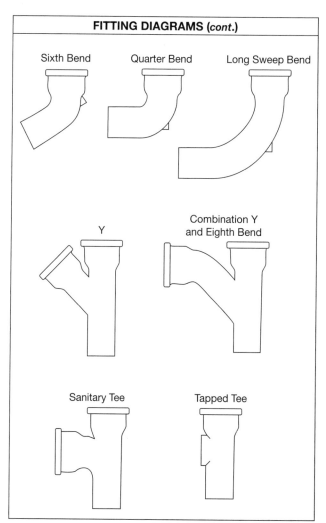

Sixth Bend

Quarter Bend

Long Sweep Bend

Y

Combination Y
and Eighth Bend

Sanitary Tee

Tapped Tee

FITTING DIAGRAMS (cont.)

Cleanout Plug

Pipe Plug

Reducer

Adapter

Closet Flange

FITTING DIAGRAMS (cont.)

P Trap

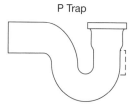

Floor Drain

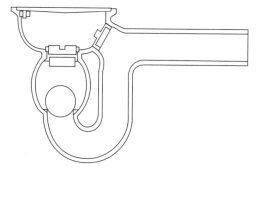

PVC SCHEDULE 40 FITTING DIMENSIONS

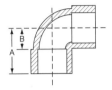

90° Elbows

Nominal Pipe Size (in.)	Center to End A (in.)	Center to Socket B (in.)	Weight (lbs.)
$\frac{1}{2}$	$1\frac{3}{8}$	$\frac{1}{2}$	.08
$\frac{3}{4}$	$1\frac{3}{4}$	$\frac{3}{4}$	.12
1	$1\frac{7}{8}$	$\frac{3}{4}$	.19
$1\frac{1}{4}$	$2\frac{3}{16}$	$\frac{15}{16}$	.28
$1\frac{1}{2}$	$2\frac{7}{16}$	$1\frac{1}{16}$	.33
2	$2\frac{13}{16}$	$1\frac{5}{16}$	.37
$2\frac{1}{2}$	$3\frac{5}{16}$	$1\frac{9}{16}$	.53
3	$3\frac{3}{4}$	$1\frac{7}{8}$	.81
4	$4\frac{5}{8}$	$2\frac{3}{8}$	1.87

Sizes may vary by manufacturer. Verify prior to installation.

PVC SCHEDULE 40 FITTING DIMENSIONS (*cont.*)

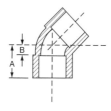

45° Elbows

Nominal Pipe Size (in.)	Center to End A (in.)	Center to Socket B (in.)	Weight (lbs.)
1/2	1 1/8	3/4	.07
3/4	1 5/16	5/16	.09
1	1 1/2	3/8	.16
1 1/4	1 11/16	7/16	.25
1 1/2	1 7/8	1/2	.30
2	2 1/8	5/8	.44
2 1/2	2 7/16	11/16	.50
3	2 11/16	13/16	.81
4	3 5/16	1 1/16	1.43

Sizes may vary by manufacturer. Verify prior to installation.

PVC SCHEDULE 40 FITTING DIMENSIONS (cont.)

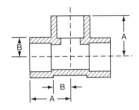

Tees

Nominal Pipe Size (in.)	Center to End A (in.)	Center to Socket B (in.)	Weight (lbs.)
1/2	1 3/8	1/2	.11
3/4	1 3/4	3/4	.17
1	1 7/8	3/4	.26
1 1/4	2 3/16	15/16	.38
1 1/2	2 7/16	1 1/16	.50
2	2 13/16	1 5/16	.72
2 1/2	3 5/16	1 9/16	1.00
3	3 3/4	1 7/8	1.37
4	4 5/8	2 3/8	2.50

Sizes may vary by manufacturer. Verify prior to installation.

PVC SCHEDULE 40 FITTING DIMENSIONS (*cont.*)

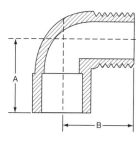

90° Street Ell

Nominal Pipe Size (in.)	Center to End A (in.)	Center to End B (in.)	Weight (lbs.)
1/2	1 3/8	1 5/8	.06
3/4	1 11/16	1 7/8	.09
1	1 7/8	2 1/8	.17
1 1/4	2 3/16	2 7/16	.25
1 1/2	2 7/8	2 11/16	.35
2	2 3/4	3 1/4	.58

Sizes may vary by manufacturer. Verify prior to installation.

PVC SCHEDULE 40 FITTING DIMENSIONS (cont.)

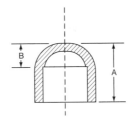

Caps

Nominal Pipe Size (in.)	End to End A (in.)	Socket to End B (in.)	Weight (lbs.)
1/2	1 1/4	3/8	.04
3/4	1 3/8	3/8	.06
1	1 1/2	3/8	.10
1 1/4	1 1/16	7/16	.16
1 1/2	1 13/16	7/16	.22
2	1 15/16	7/16	.31
2 1/2	3	1 1/4	.45
3	3 1/4	1 3/8	.52
4	3 7/8	1 5/8	.94

Sizes may vary by manufacturer. Verify prior to installation.

PVC SCHEDULE 40 FITTING DIMENSIONS (*cont.*)

Couplings

Nominal Pipe Size (in.)	End to End A (in.)	Socket to End B (in.)	Weight (lbs.)
$\frac{1}{2}$	2	$\frac{1}{4}$	.06
$\frac{3}{4}$	$2\frac{1}{4}$	$\frac{1}{4}$	.09
1	$2\frac{1}{2}$	$\frac{1}{4}$	.14
$1\frac{1}{4}$	$2\frac{3}{4}$	$\frac{1}{4}$	.24
$1\frac{1}{2}$	3	$\frac{1}{4}$	.25
2	$3\frac{1}{4}$	$\frac{1}{4}$	.44
$2\frac{1}{2}$	$3\frac{3}{4}$	$\frac{1}{4}$	.56
3	4	$\frac{1}{4}$	.63
4	$4\frac{3}{4}$	$\frac{1}{4}$	1.16

Sizes may vary by manufacturer. Verify prior to installation.

LIFT CHECK VALVE DIMENSIONS

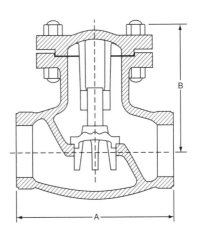

Pipe Size (in.)	Dimension A (in.)	Dimension B (in.)
$\frac{1}{2}$	$3\frac{1}{4}$	$3\frac{3}{4}$
$\frac{3}{4}$	$4\frac{1}{4}$	$3\frac{3}{4}$
1	5	$4\frac{5}{8}$
$1\frac{1}{2}$	$6\frac{1}{2}$	$5\frac{1}{2}$
2	$7\frac{1}{2}$	$5\frac{7}{8}$

Sizes may vary by manufacturer. Verify prior to installation.

SWING CHECK VALVE DIMENSIONS

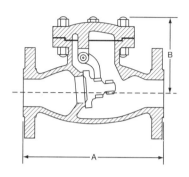

Pipe Size (in.)	Dimension A (in.)	Dimension B (in.)
$\frac{1}{2}$	6	$4\frac{1}{2}$
$\frac{3}{4}$	7	$5\frac{1}{4}$
1	$8\frac{1}{2}$	$5\frac{1}{2}$
$1\frac{1}{2}$	$9\frac{1}{2}$	5
2	$10\frac{1}{2}$	$5\frac{1}{2}$
$2\frac{1}{2}$	$11\frac{1}{2}$	6
3	$13\frac{1}{2}$	$6\frac{3}{4}$
4	14	8
6	$17\frac{1}{2}$	$9\frac{1}{2}$
8	21	$12\frac{1}{4}$
10	$24\frac{1}{2}$	14
12	28	$16\frac{1}{2}$

Sizes may vary by manufacturer. Verify prior to installation.

BRONZE GATE VALVE DIMENSIONS

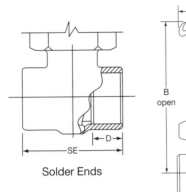

Solder Ends

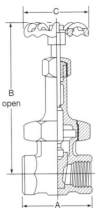

Pipe Size (in.)	SE (in.)	D (in.)	A (in.)	B (in.)	C (in.)
$\frac{1}{4}$	–	–	$1\frac{15}{16}$	$4\frac{11}{16}$	$2\frac{1}{4}$
$\frac{3}{8}$	$1\frac{21}{32}$	$\frac{3}{8}$	$1\frac{15}{16}$	$4\frac{11}{16}$	$2\frac{1}{2}$
$\frac{1}{2}$	$1\frac{29}{32}$	$\frac{1}{2}$	$2\frac{1}{4}$	$5\frac{7}{16}$	$2\frac{1}{2}$
$\frac{3}{4}$	$2\frac{1}{2}$	$\frac{3}{4}$	$2\frac{3}{8}$	$6\frac{1}{4}$	$2\frac{3}{4}$
1	$2\frac{29}{32}$	$\frac{29}{32}$	$2\frac{13}{16}$	$7\frac{1}{2}$	3
$1\frac{1}{4}$	$3\frac{3}{16}$	$\frac{11}{32}$	$3\frac{1}{8}$	$8\frac{5}{8}$	$3\frac{1}{2}$
$1\frac{1}{2}$	$3\frac{5}{8}$	$1\frac{3}{32}$	$3\frac{3}{8}$	10	$3\frac{23}{32}$
2	$4\frac{1}{4}$	$1\frac{11}{32}$	$3\frac{1}{2}$	$11\frac{5}{8}$	$4\frac{1}{32}$
$2\frac{1}{2}$	$4\frac{15}{16}$	$1\frac{5}{8}$	$4\frac{5}{16}$	15	5
3	$5\frac{5}{8}$	$1\frac{7}{8}$	$4\frac{9}{16}$	$17\frac{1}{2}$	5

Sizes may vary by manufacturer. Verify prior to installation.

ANGLE AND GLOBE VALVE DIMENSIONS

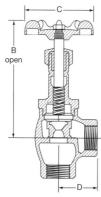

Angle Globe

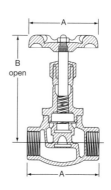

Globe Valve

Pipe Size (in.)	A (in.)	B (in.)	C (in.)	D (in.)
$\frac{1}{8}$	$2\frac{3}{8}$	$3\frac{3}{8}$	$2\frac{3}{8}$	–
$\frac{1}{4}$	$2\frac{3}{8}$	$3\frac{3}{8}$	$2\frac{3}{8}$	$1\frac{3}{16}$
$\frac{3}{8}$	$2\frac{3}{8}$	$3\frac{3}{8}$	$2\frac{3}{8}$	$1\frac{3}{16}$
$\frac{1}{2}$	$2\frac{9}{16}$	$3\frac{3}{8}$	$2\frac{3}{8}$	$1\frac{5}{16}$
$\frac{3}{4}$	$3\frac{1}{16}$	$4\frac{3}{4}$	$2\frac{3}{4}$	$1\frac{9}{16}$
1	$3\frac{11}{16}$	$5\frac{11}{16}$	$2\frac{3}{4}$	$1\frac{7}{8}$
$1\frac{1}{4}$	$4\frac{5}{16}$	$6\frac{1}{8}$	3	$2\frac{3}{16}$
$1\frac{1}{2}$	$4\frac{11}{16}$	$7\frac{3}{16}$	$3\frac{11}{16}$	$2\frac{3}{8}$
2	$5\frac{5}{8}$	$7\frac{15}{16}$	$4\frac{1}{32}$	$2\frac{13}{16}$
$2\frac{1}{2}$	$6\frac{5}{8}$	$10\frac{3}{16}$	5	$3\frac{3}{16}$
3	$7\frac{3}{4}$	$11\frac{3}{16}$	6	$3\frac{7}{8}$

Sizes may vary by manufacturer. Verify prior to installation.

DIMENSIONS OF FERROUS VALVES IN INCHES

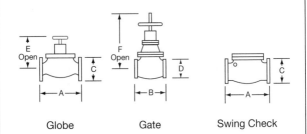

| Globe | Gate | Swing Check |

Pipe	Maximum rating 150 psi					
Size	A	B	C	D	E	F
2	8	7	8	6	$13\frac{3}{4}$	$16\frac{1}{4}$
$2\frac{1}{2}$	$8\frac{1}{2}$	$7\frac{1}{2}$	$8\frac{1}{2}$	7	$14\frac{1}{2}$	$17\frac{1}{2}$
3	$9\frac{1}{2}$	8	$9\frac{1}{2}$	$7\frac{1}{2}$	$16\frac{1}{2}$	21
4	$11\frac{1}{2}$	9	$11\frac{1}{2}$	9	$19\frac{3}{4}$	26
5	14	10	13	10	23	31
6	16	$10\frac{1}{2}$	14	11	$24\frac{1}{2}$	$34\frac{1}{2}$
8	$19\frac{1}{2}$	$11\frac{1}{2}$	$19\frac{1}{2}$	$13\frac{1}{2}$	26	$42\frac{1}{2}$
10	–	13	$24\frac{1}{2}$	16	–	$51\frac{1}{2}$
12	–	14	–	19	–	$59\frac{1}{4}$
14	–	15	–	21	–	$70\frac{1}{4}$

Sizes may vary by manufacturer. Verify prior to installation.

DIMENSIONS OF FERROUS VALVES IN INCHES (*cont.*)

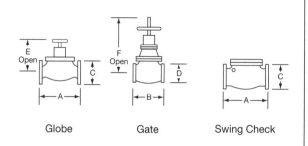

Globe   Gate   Swing Check

Pipe Size	Maximum rating 300 psi					
	A	B	C	D	E	F
2	$10\frac{1}{2}$	$8\frac{1}{2}$	$10\frac{1}{2}$	$6\frac{1}{2}$	$17\frac{3}{4}$	$18\frac{1}{4}$
$2\frac{1}{2}$	$11\frac{1}{2}$	$9\frac{1}{2}$	$11\frac{1}{2}$	$7\frac{1}{2}$	19	$21\frac{1}{4}$
3	$12\frac{1}{2}$	$11\frac{1}{8}$	$12\frac{1}{2}$	$8\frac{1}{4}$	$20\frac{1}{2}$	25
4	14	12	14	10	$24\frac{3}{4}$	31
5	$15\frac{3}{4}$	15	$15\frac{3}{4}$	11	$26\frac{1}{2}$	$34\frac{1}{2}$
6	$17\frac{1}{2}$	$15\frac{7}{8}$	$17\frac{1}{2}$	$12\frac{1}{2}$	$29\frac{3}{4}$	$38\frac{1}{2}$
8	22	$16\frac{1}{2}$	21	15	$35\frac{1}{2}$	48
10	$24\frac{1}{2}$	18	$24\frac{1}{2}$	$17\frac{1}{2}$	–	59
12	–	$19\frac{3}{4}$	28	$20\frac{1}{2}$	–	$66\frac{1}{4}$
14	–	$22\frac{1}{2}$	–	23	–	$74\frac{3}{4}$

Sizes may vary by manufacturer. Verify prior to installation.

DIMENSIONS OF FERROUS VALVES IN INCHES (*cont.*)

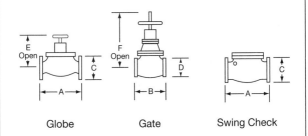

Globe　　　　　Gate　　　　　Swing Check

Pipe Size	Maximum rating 400 psi					
	A	B	C	D	E	F
2	$11\frac{1}{2}$	$11\frac{1}{2}$	$11\frac{1}{2}$	$6\frac{1}{2}$	21	$20\frac{1}{4}$
$2\frac{1}{2}$	13	13	13	$7\frac{1}{2}$	$23\frac{1}{8}$	$22\frac{1}{8}$
3	14	14	14	$8\frac{1}{4}$	$25\frac{1}{8}$	$24\frac{5}{8}$
4	16	16	16	10	$28\frac{7}{8}$	$29\frac{1}{8}$
5	18	18	18	11	–	$35\frac{1}{8}$
6	$19\frac{1}{2}$	$19\frac{1}{2}$	$19\frac{1}{2}$	$12\frac{1}{2}$	36	$38\frac{3}{4}$
8	$23\frac{1}{2}$	$23\frac{1}{2}$	$23\frac{1}{2}$	15	48	$48\frac{5}{8}$
10	$26\frac{1}{2}$	$26\frac{1}{2}$	$26\frac{1}{2}$	$17\frac{1}{2}$	$58\frac{1}{4}$	$58\frac{3}{4}$
12	–	30	30	$20\frac{1}{2}$	–	66
14	–	$33\frac{1}{2}$	–	23	–	–

Sizes may vary by manufacturer. Verify prior to installation.

DIMENSIONS OF FERROUS VALVES IN INCHES (*cont.*)

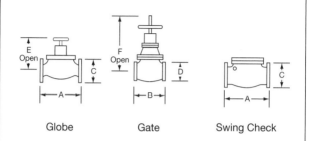

Globe   Gate   Swing Check

Pipe Size	Maximum rating 600 psi					
	A	B	C	D	E	F
2	$11\frac{1}{2}$	$11\frac{1}{2}$	$11\frac{1}{2}$	$6\frac{1}{2}$	19	$18\frac{1}{4}$
$2\frac{1}{2}$	13	13	13	$7\frac{1}{2}$	$21\frac{1}{4}$	$22\frac{1}{4}$
3	14	14	14	$8\frac{1}{4}$	$23\frac{1}{2}$	$25\frac{3}{4}$
4	17	17	17	$10\frac{3}{4}$	$27\frac{1}{2}$	32
5	20	20	20	13	$30\frac{3}{8}$	$36\frac{3}{4}$
6	22	22	22	14	35	$42\frac{3}{4}$
8	26	26	26	$16\frac{1}{2}$	$45\frac{3}{8}$	$52\frac{1}{4}$
10	31	31	31	20	54	$62\frac{1}{4}$
12	–	33	33	22	–	70
14	–	–	–	$23\frac{3}{4}$	–	$77\frac{1}{4}$

Sizes may vary by manufacturer. Verify prior to installation.

QUARTER BEND AND SANITARY T-BRANCH
FITTING DIMENSIONS AND WEIGHTS

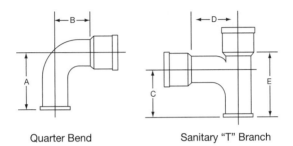

Quarter Bend   Sanitary "T" Branch

Pipe Size (in.)	Quarter Bend			Sanitary T			
	A (in.)	B (in.)	Wt. (lbs.)	C (in.)	D (in.)	E (in.)	Wt. (lbs.)
1½	5¾	3¼	6	6¾	3¼	8⅜	9
2	6	3½	7	7	3½	9	12
3	7	4	12	7½	4	10	17
4	8	4½	17	8	5	11	25
5	8½	5	21	8½	5	12	31
6	9	5½	31	9	5½	13	47

Sizes may vary by manufacturer. Verify prior to installation.

WELDING FITTING DIMENSIONS IN INCHES

Pipe Size	O.D.	45° Elbow Thickness	Straight Tee A	Eccentric B	Concentric C	D	Cap E
¾	1.050	.083	1⅛	⁷⁄₁₆	1⅛	–	–
1	1.315	.109	1½	⅞	1½	2	1½
1¼	1.660	.109	1⅞	1	1⅞	2	1½
1½	1.900	.109	2¼	1⅛	2¼	2½	1½
2	2.375	.109	3	1⅜	2½	3	1½
2½	2.875	.120	3¾	1¾	3	3½	1½
3	3.500	.120	4½	2	3⅜	3½	2
3½	4.000	.120	5¼	2¼	3¾	4	2½
4	4.500	.120	6	2½	4⅛	4	2½
6	6.625	.134	9	3¾	5⅝	5½	3½
8	8.625	.148	12	5	7	6	4
10	10.750	.165	15	6¼	8½	7	5
12	12.750	.180	18	7½	10	8	6

Sizes may vary by manufacturer. Verify prior to installation.

90° Elbow   45° Elbow   Straight Tee   Eccentric   Concentric   Cap

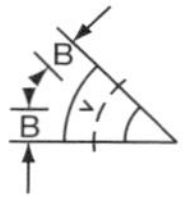

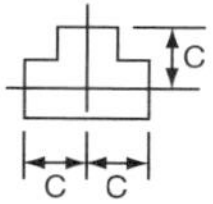

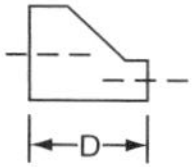

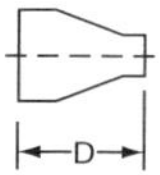

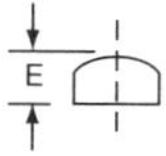

STEEL-WELDED FITTING DIMENSIONS IN INCHES

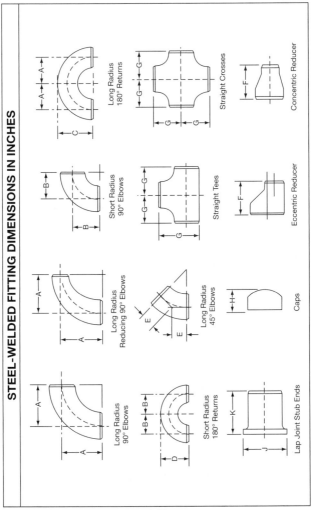

Long Radius 90° Elbows

Long Radius Reducing 90° Elbows

Short Radius 90° Elbows

Long Radius 180° Returns

Long Radius 45° Elbows

Straight Tees

Straight Crosses

Short Radius 180° Returns

Lap Joint Stub Ends

Caps

Eccentric Reducer

Concentric Reducer

STEEL-WELDED FITTING DIMENSIONS IN INCHES (cont.)

Nominal Pipe Size	A	B	C	D	E	F	G	H	J	K
1	1½	1	2$\frac{1}{16}$	1$\frac{5}{8}$	$\frac{7}{8}$	2	1½	1½	2	4
1½	2¼	1½	3¼	2$\frac{7}{16}$	1$\frac{1}{8}$	2½	2¼	1½	2$\frac{7}{8}$	4
2	3	2	4$\frac{3}{16}$	3$\frac{5}{16}$	1$\frac{3}{8}$	3	2½	1½	3$\frac{5}{8}$	6
3	4½	3	6¼	4¾	2	3½	3$\frac{3}{8}$	2	5	6
4	6	4	8¼	6¼	2½	4	4$\frac{1}{8}$	2½	6$\frac{3}{16}$	6
6	9	6	12¾	9$\frac{5}{16}$	3¾	5½	5$\frac{5}{8}$	3½	8½	8
8	12	8	16$\frac{5}{16}$	12$\frac{5}{16}$	5	6	7	4	10$\frac{5}{8}$	8
10	15	10	20$\frac{3}{8}$	15$\frac{3}{8}$	6¼	7	8½	5	12¾	10
12	18	12	24$\frac{3}{8}$	18$\frac{3}{8}$	7½	8	10	6	15	10
14	21	14	28	21	8¾	13	11	6½	16¼	12
16	24	16	32	23	10	14	12	7	18½	12
18	27	18	36	25½	11¼	15	13½	8	21	12
20	30	20	40	30½	12½	20	15	9	23	12
24	36	24	48	34	15	20	17	10½	27¼	12

Sizes may vary by manufacturer. Verify prior to installation.

STAINLESS STEEL FITTING DIMENSIONS IN INCHES

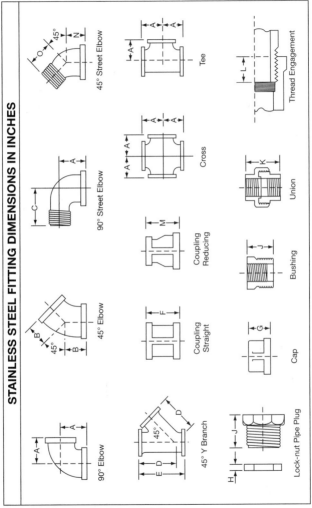

90° Elbow

45° Elbow

90° Street Elbow

45° Street Elbow

45° Y Branch

Coupling Straight

Coupling Reducing

Tee

Cross

Lock-nut Pipe Plug

Cap

Bushing

Union

Thread Engagement

STAINLESS STEEL FITTING DIMENSIONS IN INCHES (cont.)

Pipe Size	A	B	C	D	E	F	G	H	J	K	L	M	N	O
1/8	11/16	11/16	1	1	15/16	7/8	5/8	3/16	5/8	13/16	1/4	—	—	—
1/4	13/16	3/4	13/16	13/16	15/8	11/16	11/16	1/4	11/16	11/2	3/8	1	3/4	15/16
3/8	15/16	13/16	17/16	17/16	115/16	13/16	11/16	9/32	3/4	15/8	3/8	11/8	13/16	11/32
1/2	11/8	7/8	15/8	111/16	25/16	15/16	7/8	5/16	15/16	111/16	7/16	11/4	7/8	15/32
3/4	15/16	1	17/8	21/16	23/4	11/2	1	11/32	1	115/16	1/2	17/16	1	15/16
1	11/2	11/8	21/8	27/16	35/16	111/16	13/16	3/8	11/8	21/2	9/16	111/16	11/8	11/2
11/4	13/4	15/16	27/16	215/16	315/16	115/16	19/32	7/16	11/8	3	11/16	21/16	15/16	123/32
11/2	115/16	17/16	211/16	35/16	43/8	21/8	15/16	15/32	11/8	3	3/4	25/16	17/16	17/8
2	21/4	111/16	31/4	315/16	53/16	21/2	17/16	17/32	11/4	33/8	3/4	213/16	111/16	27/32
21/2	211/16	115/16	37/8	43/4	61/4	27/8	111/16	19/32	17/16	31/2	7/8	31/4	115/16	29/16
3	31/16	23/16	41/2	59/16	71/4	33/16	113/16	11/16	11/2	41/8	1	311/16	23/16	3
4	313/16	25/8	511/16	7	9	311/16	21/16	13/16	13/4	43/8	11/16	43/8	25/8	311/16

Sizes may vary by manufacturer. Verify prior to installation.

STAINLESS STEEL FLANGED FITTING DIMENSIONS IN INCHES

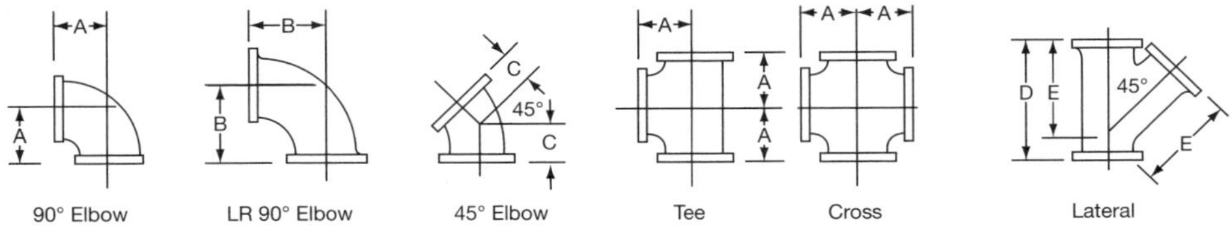

90° Elbow | LR 90° Elbow | 45° Elbow | Tee | Cross | Lateral

Flange Dimensions

Pipe Size	Flange Diameter	Minimum Thickness	Raised Face
1	4¼	⅜	2
1¼	4⅝	13/32	2½
1½	5	7/16	2⅞
2	6	½	3⅝
2½	7	9/16	4⅛
3	7½	⅝	5
4	9	11/16	6³/16
6	11	13/16	8½
8	13½	15/16	10⅝
10	16	1	12¾

Center-to-Face Dimensions

A	B	C	D	E
3½	5	1¾	7½	5¾
3¾	5½	2	8	6¼
4	6	2¼	9	7
4½	6½	2½	10½	8
5	7	3	12	9½
5½	7¾	3	13	10
6½	9	4	15	12
8	11½	5	18	14½
9	14	5½	22	17½
11	16½	5½	25½	20½

Sizes may vary by manufacturer. Verify prior to installation.

LIGHTWEIGHT FLANGE DIMENSIONS IN INCHES

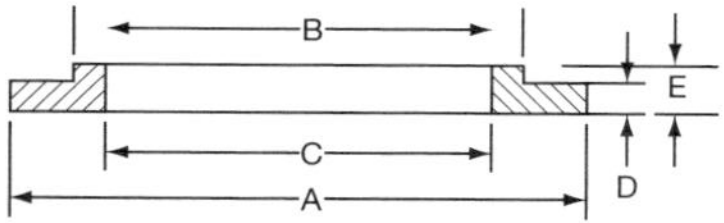

Pipe Size	O.D. A	O.D. B	I.D. C	Thickness D	Hub Length E	Number of Bolts
3	7½	4¼	3¼	½	⅞	4
4	9	5⁵⁄₁₆	4¼	½	⅞	8
5	10	6⁷⁄₁₆	5¼	⁹⁄₁₆	⅞	8
6	11	7⁹⁄₁₆	6¼	⁹⁄₁₆	1⅛	8
8	13½	9¹¹⁄₁₆	8¼	⁹⁄₁₆	1⅛	8
10	16	12	10⁵⁄₁₆	¹¹⁄₁₆	1³⁄₁₆	12
12	19	14⅜	12⁵⁄₁₆	¹¹⁄₁₆	1⁷⁄₁₆	12
14	21	15¾	14⁵⁄₁₆	¾	1⁷⁄₁₆	12
16	23½	18	16⁵⁄₁₆	¾	1½	16
18	25	19⅜	18³⁄₈	¾	1½	16
20	27½	22	20⅜	¾	1½	20
22	29½	24⅛	22⁷⁄₁₆	1	1½	20
24	32	26⅛	24⁷⁄₁₆	1	1⅞	20
26	34¼	28½	26⁹⁄₁₆	1	1⅞	24
28	36½	30½	28⁹⁄₁₆	1	1⅞	28

Sizes may vary by manufacturer. Verify prior to installation.

CAST IRON FLANGE DIMENSIONS IN INCHES

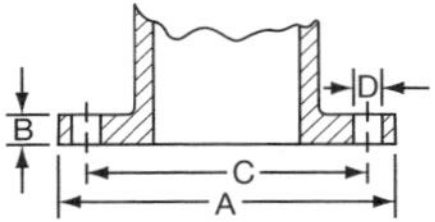

Class 125 Cast Iron

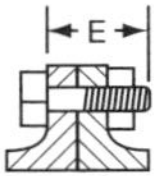

Length of Machine Bolt

Nominal Pipe Size	Flanges		Drilling		Bolting		
	Flange Diameter (A)	Flange Thickness (B)	Diameter of Bolt Circle (C)	Diameter of Bolt Hole (D)	Diameter of Bolt	Length of Machine Bolt (E)	Number of Bolts
1	4 1/4	7/16	3 1/8	5/8	1/2	1 3/4	4
1 1/4	4 5/8	1/2	3 1/2	5/8	1/2	2	4
1 1/2	5	5/16	3 7/8	5/8	1/2	2	4
2	6	5/8	4 3/4	3/4	5/8	2 1/4	4
2 1/2	7	11/16	5 1/2	3/4	5/8	2 1/2	4
3	7 1/2	3/4	6	3/4	5/8	2 1/2	4
4	9	15/16	7 1/2	3/4	5/8	3	8
5	10	15/16	8 1/2	7/8	3/4	3	8
6	11	1	9 1/2	7/8	3/4	3 1/4	8
8	13 1/2	1 1/8	11 3/4	7/8	3/4	3 1/2	8

Sizes may vary by manufacturer. Verify prior to installation.

CAST IRON FLANGE DIMENSIONS IN INCHES (cont.)

Class 125 Cast Iron

Length of Machine Bolt

Nominal Pipe Size	Flanges		Drilling		Bolting		
	Flange Diameter (A)	Flange Thickness (B)	Diameter of Bolt Circle (C)	Diameter of Bolt Hole (D)	Diameter of Bolt	Length of Machine Bolt (E)	Number of Bolts
10	16	1 1/4	14 1/4	1	7/8	3 3/4	12
12	19	1 1/4	17	1	7/8	3 3/4	12
14	21	1 3/8	18 3/4	1 1/8	1	4 1/4	12
16	23 1/2	1 7/16	21 1/4	1 1/8	1	4 1/2	16
18	25	1 9/16	22 3/4	1 1/4	1 1/8	4 3/4	16
20	27 1/2	1 11/16	25	1 1/4	1 1/8	5	20
24	32	1 7/8	29 1/2	1 3/8	1 1/4	5 1/2	20
30	38 3/4	2 1/8	36	1 3/8	1 1/4	6 1/4	28
36	46	2 3/8	42 3/4	1 5/8	1 1/2	7	32

Sizes may vary by manufacturer. Verify prior to installation.

1-45

IRON SOIL PIPE FITTING DIMENSIONS IN INCHES

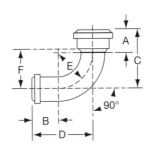

Short Sweep Elbows

Pipe Size	A	B	C	D	E	F	Telescoping Length
2	2³⁄₄	3	7³⁄₄	8	5	5¹⁄₄	2¹⁄₂
3	3¹⁄₄	3¹⁄₂	8³⁄₄	9	5¹⁄₂	6	2³⁄₄
4	3¹⁄₂	4	9¹⁄₈	10	6	6¹⁄₂	3
5	3¹⁄₂	4	10	10¹⁄₂	6¹⁄₂	7	3
6	3¹⁄₂	4	10¹⁄₂	11	7	7¹⁄₂	3
8	4¹⁄₈	5¹⁄₂	12¹⁄₈	13¹⁄₂	8	9⁵⁄₈	3¹⁄₂
10	4¹⁄₈	5¹⁄₂	13¹⁄₈	14¹⁄₂	9	9⁵⁄₈	3¹⁄₂
12	5	7	15	17	10	10³⁄₄	4¹⁄₄
15	5	7	16¹⁄₂	18¹⁄₂	11¹⁄₂	12¹⁄₄	4¹⁄₄

Sizes may vary by manufacturer. Verify prior to installation.

IRON SOIL PIPE FITTING DIMENSIONS IN INCHES (*cont.*)

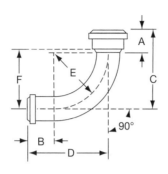

Long Sweep Elbows

Pipe Size	A	B	C	D	E	F	Telescoping Length
2	2¾	3	10¾	11	8	8¼	2½
3	3¼	3½	11¾	12	8½	9	2¾
4	3½	4	12½	13	9	9½	3
5	3½	4	13	13½	9½	10	3
6	3½	4	13½	14	10	10½	3
8	4⅛	5½	15⅛	16½	11	11⅝	3½
10	4⅛	5½	16⅛	17½	12	12⅝	3½
12	5	7	18	20	13	13¾	4¼
15	5	7	19½	21½	14½	15¼	4¼

Sizes may vary by manufacturer. Verify prior to installation.

IRON SOIL PIPE FITTING DIMENSIONS IN INCHES (cont.)

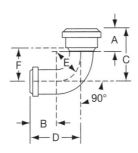

¼ Bend Elbows

Pipe Size	A	B	C	D	E	F	Telescoping Length
2	2¾	3	5¾	6	3	3¼	2½
3	3¼	3½	6¾	7	3½	4	2¾
4	3½	4	7½	8	4	4½	3
5	3½	4	8	8½	4½	5	3
6	3½	4	8½	9	5	5½	3
8	4⅛	5½	10⅛	11½	6	6⅝	3½
10	4⅛	5½	11⅛	12½	7	7⅝	3½
12	5	7	13	15	8	8¾	4¼
15	5	7	14½	16½	9½	10¼	4¼

Sizes may vary by manufacturer. Verify prior to installation.

IRON SOIL PIPE FITTING DIMENSIONS IN INCHES (*cont.*)

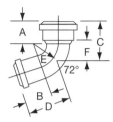

⅛ Bend Elbows

Pipe Size	A	B	C	D	E	F	Telescoping Length
2	2³⁄₄	3	4	4¹⁄₄	3	1¹⁄₂	2¹⁄₂
3	3¹⁄₄	3¹⁄₂	4¹¹⁄₁₆	4¹⁵⁄₁₆	3¹⁄₂	1¹⁵⁄₁₆	2³⁄₄
4	3¹⁄₂	4	5³⁄₁₆	5¹¹⁄₁₆	4	2³⁄₁₆	3
5	3¹⁄₂	4	5³⁄₈	5⁷⁄₈	4¹⁄₂	2³⁄₈	3
6	3¹⁄₂	4	5⁹⁄₁₆	6¹⁄₁₆	5	2⁹⁄₁₆	3
8	4¹⁄₈	5¹⁄₂	6⁵⁄₈	8	6	3¹⁄₈	3¹⁄₂
10	4¹⁄₈	5¹⁄₂	7	8³⁄₈	7	3¹⁄₂	3¹⁄₂
12	5	7	8⁵⁄₁₆	10⁵⁄₁₆	8	4¹⁄₁₆	4¹⁄₄
15	5	7	8¹⁵⁄₁₆	10¹⁵⁄₁₆	9¹⁄₂	4¹¹⁄₁₆	4¹⁄₄

Sizes may vary by manufacturer. Verify prior to installation.

IRON SOIL PIPE FITTING DIMENSIONS IN INCHES (*cont.*)

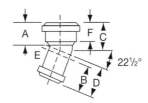

¹⁄₁₆ Bend Elbows

Pipe Size	A	B	C	D	E	F	Telescoping Length
2	$2\frac{3}{4}$	3	$3\frac{3}{8}$	$3\frac{5}{8}$	3	$\frac{7}{8}$	$2\frac{1}{2}$
3	$3\frac{1}{4}$	$3\frac{1}{2}$	$3\frac{15}{16}$	$4\frac{3}{16}$	$3\frac{1}{2}$	$1\frac{3}{16}$	$2\frac{3}{4}$
4	$3\frac{1}{2}$	4	$4\frac{5}{16}$	$4\frac{13}{16}$	4	$1\frac{5}{16}$	3
5	$3\frac{1}{2}$	4	$4\frac{3}{8}$	$4\frac{7}{8}$	$4\frac{1}{2}$	$1\frac{3}{8}$	3
6	$3\frac{1}{2}$	4	$4\frac{1}{2}$	5	5	$1\frac{1}{2}$	3
8	$4\frac{1}{8}$	$5\frac{1}{2}$	$5\frac{5}{16}$	$6\frac{11}{16}$	6	$1\frac{13}{16}$	$3\frac{1}{2}$
10	$4\frac{1}{8}$	$5\frac{1}{2}$	$5\frac{1}{2}$	$6\frac{7}{8}$	7	2	$3\frac{1}{2}$
12	5	7	$6\frac{5}{8}$	$8\frac{5}{8}$	8	$2\frac{3}{8}$	$4\frac{1}{4}$
15	5	7	$6\frac{7}{8}$	$8\frac{7}{8}$	$9\frac{1}{2}$	$2\frac{5}{8}$	$4\frac{1}{4}$

Sizes may vary by manufacturer. Verify prior to installation.

CHAPTER 2
Pipefitting

MEASURING LENGTH OF PIPE IN BENDS

The length of pipe (arc) in any bend depends on the following:

(A) Number of degrees in angle of bend

(B) Length of bending radius

The length of arc in a pipe bend is measured along the centerline of the pipe. The radius is also measured as extending to the centerline. Use the formula below to calculate arc length.

Length of Arc = No. of Degrees X .0175 X Radius of Bend

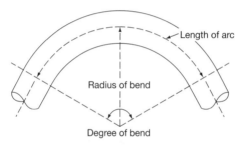

Length of arc

Radius of bend

Degree of bend

FINDING LENGTH OF PIPE IN STANDARD BENDS

If Angle in Degrees is	Multiply Radius by	If Angle in Degrees is	Multiply Radius by
22½	.393	150	2.618
30	.524	180	3.142
45	.785	210	3.665
60	1.047	240	4.189
90	1.571	270	4.712
112½	1.963	300	5.236
120	2.094	360	6.283
135	2.356	540	9.425

SETBACK AND LENGTH OF PIPE BENDS

The setback is a measurement used to locate the beginning of a bend in a pipe. There is a setback for each angle and radius of bend.

Setback is the distance from the beginning of the bend to the point where extended centerlines of straight pipe on each leg of the bend would meet.

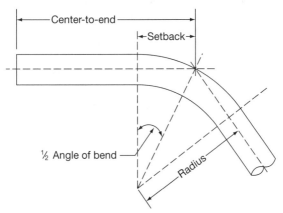

Tables of values for setback and length of bend at standard angles and various radii can be found in this chapter. In cases not covered by these tables, the values can be calculated by using trigonometry. The setback is calculated as follows:

Setback = Radius X Tangent ½ Angle of Bend

The values of Tangent ½ Angle of Bend for common angles are:

22½°	.199
30°	.268
45°	.414
60°	.597
90°	1.000

SETBACK AND LENGTH OF 22½° BENDS

The setback and length of bend for 22½° pipe bends are calculated by using these formulas:

Setback = Radius X .199

Length of Bend = Radius X .393

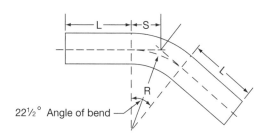

22½° Angle of bend

L = Straight Pipe Length
R = Bend Radius Length
S = Setback Length

SETBACK AND LENGTH OF 22½° BENDS WITH RADII FROM 1 TO 72 INCHES

Radius of Bend (in.)	Setback (in.)	Length of Bend (in.)
1	3/16	3/8
2	3/8	13/16
3	5/8	1 3/16
4	13/16	1 9/16
5	1	2
6	1 3/16	2 3/8
7	1 3/8	2 3/4
8	1 5/8	3 1/8
9	1 13/16	3 1/2
10	2	3 15/16

2-3

SETBACK AND LENGTH OF 22½° BENDS WITH RADII FROM 1 TO 72 INCHES (cont.)

Radius of Bend (in.)	Setback (in.)	Length of Bend (in.)
11	$2\frac{3}{16}$	$4\frac{5}{16}$
12	$2\frac{3}{8}$	$4\frac{11}{16}$
13	$2\frac{5}{8}$	$5\frac{1}{8}$
14	$2\frac{13}{16}$	$5\frac{1}{2}$
15	3	$5\frac{13}{16}$
16	$3\frac{3}{16}$	$6\frac{1}{4}$
17	$3\frac{3}{8}$	$6\frac{11}{16}$
18	$3\frac{5}{8}$	$7\frac{1}{16}$
19	$3\frac{13}{16}$	$7\frac{7}{16}$
20	4	$7\frac{7}{8}$
21	$4\frac{3}{16}$	$8\frac{1}{4}$
22	$4\frac{3}{8}$	$8\frac{5}{8}$
23	$4\frac{5}{8}$	$9\frac{1}{16}$
24	$4\frac{13}{16}$	$9\frac{7}{16}$
25	5	$9\frac{13}{16}$
26	$5\frac{3}{16}$	$10\frac{3}{16}$
27	$5\frac{3}{8}$	$10\frac{5}{8}$
28	$5\frac{5}{8}$	11
29	$5\frac{13}{16}$	$11\frac{3}{8}$
30	6	$11\frac{7}{8}$
31	$6\frac{3}{16}$	$12\frac{3}{16}$
32	$6\frac{1}{2}$	$12\frac{9}{16}$
33	$6\frac{9}{16}$	$12\frac{15}{16}$
34	$6\frac{3}{4}$	$13\frac{3}{8}$
35	7	$13\frac{3}{4}$
36	$7\frac{3}{16}$	$14\frac{1}{8}$
37	$7\frac{3}{8}$	$14\frac{1}{2}$
38	$7\frac{9}{16}$	$14\frac{15}{16}$
39	$7\frac{3}{4}$	$15\frac{5}{16}$
40	8	$15\frac{3}{4}$
41	$8\frac{3}{16}$	$16\frac{1}{8}$

SETBACK AND LENGTH OF 22½° BENDS WITH RADII FROM 1 TO 72 INCHES (*cont.*)

Radius of Bend (in.)	Setback (in.)	Length of Bend (in.)
42	8⅜	16½
43	8⁹⁄₁₆	16⅞
44	8¾	17¼
45	9	17⅝
46	9⅛	18¹⁄₁₆
47	9⅜	18½
48	9⁹⁄₁₆	18⅞
49	9¾	19¼
50	10	19¹¹⁄₁₆
51	10⅛	20
52	10⅜	20⁷⁄₁₆
53	10½	20¹³⁄₁₆
54	10¾	21³⁄₁₆
55	10¹⁵⁄₁₆	21⅝
56	11⅛	22
57	11⁵⁄₁₆	22⅜
58	11⁹⁄₁₆	22¾
59	11¹³⁄₁₆	23⅛
60	11⅞	23⁹⁄₁₆
61	12³⁄₁₆	24
62	12⅜	24⅜
63	12⅝	24¹³⁄₁₆
64	12¹³⁄₁₆	25³⁄₁₆
65	13	25⅝
66	13³⁄₁₆	26
67	13⅜	26⅜
68	13⅝	26¹³⁄₁₆
69	13¹³⁄₁₆	27³⁄₁₆
70	14	27⁹⁄₁₆
71	14³⁄₁₆	27¹⁵⁄₁₆
72	14⅜	28⅜

SETBACK AND LENGTH OF 30° BENDS

The setback and length of bend for 30° pipe bends are calculated by using these formulas:

Setback = Radius X .268

Length of Bend = Radius X .524

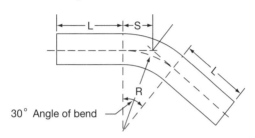

30° Angle of bend

L = Straight Pipe Length
R = Bend Radius Length
S = Setback Length

SETBACK AND LENGTH OF 30° BENDS WITH RADII FROM 1 TO 72 INCHES

Radius of Bend (in.)	Setback (in.)	Length of Bend (in.)
1	¼	½
2	⁹⁄₁₆	1¹⁄₁₆
3	¹³⁄₁₆	1⁹⁄₁₆
4	1¹⁄₁₆	2⅛
5	1⅜	2⅝
6	1⅝	3⅛
7	1⅞	3¹¹⁄₁₆
8	2⅛	4⅛
9	2⁷⁄₁₆	4¹¹⁄₁₆
10	2¹¹⁄₁₆	5¼

Radius of Bend (in.)	Setback (in.)	Length of Bend (in.)
11	$2^{15}/_{16}$	$5^3/_4$
12	$3^1/_4$	$6^5/_{16}$
13	$3^1/_2$	$6^{13}/_{16}$
14	$3^3/_4$	$7^3/_8$
15	4	$7^7/_8$
16	$4^5/_{16}$	$8^3/_8$
17	$4^9/_{16}$	$8^{15}/_{16}$
18	$4^{13}/_{16}$	$9^7/_{16}$
19	$5^1/_8$	10
20	$5^3/_8$	$10^1/_2$
21	$5^5/_8$	11
22	$5^7/_8$	$11^9/_{16}$
23	$6^3/_{16}$	$12^1/_{16}$
24	$6^7/_{16}$	$12^5/_8$
25	$6^{11}/_{16}$	$13^1/_8$
26	7	$13^5/_8$
27	$7^1/_4$	$14^3/_{16}$
28	$7^1/_2$	$14^{11}/_{16}$
29	$7^3/_4$	$15^3/_{16}$
30	$8^1/_{16}$	$15^3/_4$
31	$8^5/_{16}$	$16^1/_4$
32	$8^9/_{16}$	$16^{13}/_{16}$
33	$8^7/_8$	$17^3/_{16}$
34	$9^1/_8$	$17^{13}/_{16}$
35	$9^3/_8$	$18^3/_8$
36	$9^{11}/_{16}$	$18^7/_8$
37	$9^{15}/_{16}$	$19^3/_8$
38	$10^3/_{16}$	20
39	$10^1/_2$	$20^7/_{16}$
40	$10^3/_4$	21
41	11	$21^1/_2$

SETBACK AND LENGTH OF 30° BENDS WITH RADII FROM 1 TO 72 INCHES (cont.)

Radius of Bend (in.)	Setback (in.)	Length of Bend (in.)
42	11¼	22
43	11½	22½
44	11¹³⁄₁₆	23¹⁄₁₆
45	12¹⁄₁₆	23⁹⁄₁₆
46	12⅜	24⅛
47	12⅝	24⅝
48	12⅞	25⅛
49	13⅛	25¹¹⁄₁₆
50	13⁷⁄₁₆	26³⁄₁₆
51	13¹¹⁄₁₆	26¾
52	13⁵⁄₁₆	27¼
53	14¼	27¾
54	14½	28⁵⁄₁₆
55	14¾	28¹³⁄₁₆
56	15	29⅜
57	15¼	29⅞
58	15⁹⁄₁₆	30⅜
59	15¾	30⅞
60	16⅛	31⁷⁄₁₆
61	16⅜	32
62	16⅝	32½
63	16⅞	33¹⁄₁₆
64	17³⁄₁₆	33⅝
65	17⁷⁄₁₆	34⅛
66	17¹¹⁄₁₆	34¹¹⁄₁₆
67	18	35³⁄₁₆
68	18¼	35⁹⁄₁₆
69	18½	36¼
70	18¾	36¾
71	19¹⁄₁₆	37¼
72	19⁵⁄₁₆	37¹³⁄₁₆

SETBACK AND LENGTH OF 45° BENDS

The setback and length of bend for 45° pipe bends are calculated by using these formulas:

Setback = Radius X .414

Length of Bend = Radius X .785

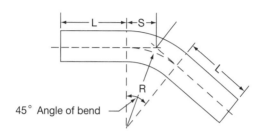

45° Angle of bend

L = Straight Pipe Length
R = Bend Radius Length
S = Setback Length

SETBACK AND LENGTH OF 45° BENDS WITH RADII FROM 1 TO 72 INCHES

Radius of Bend (in.)	Setback (in.)	Length of Bend (in.)
1	$7/16$	$3/4$
2	$13/16$	$1\,9/16$
3	$1\,1/4$	$2\,3/8$
4	$1\,5/8$	$3\,1/8$
5	$2\,1/16$	$3\,15/16$
6	$2\,7/16$	$4\,11/16$
7	$2\,7/8$	$5\,1/2$
8	$3\,1/4$	$6\,1/4$
9	$3\,3/4$	$7\,1/16$
10	$4\,1/8$	$7\,13/16$

SETBACK AND LENGTH OF 45° BENDS WITH RADII FROM 1 TO 72 INCHES (cont.)

Radius of Bend (in.)	Setback (in.)	Length of Bend (in.)
11	$4\frac{1}{2}$	$8\frac{5}{8}$
12	$4\frac{15}{16}$	$9\frac{7}{16}$
13	$5\frac{5}{16}$	$10\frac{3}{16}$
14	$5\frac{3}{4}$	11
15	$6\frac{1}{8}$	$11\frac{3}{4}$
16	$6\frac{9}{16}$	$12\frac{9}{16}$
17	7	$13\frac{5}{16}$
18	$7\frac{7}{16}$	$14\frac{1}{16}$
19	$7\frac{7}{8}$	$14\frac{15}{16}$
20	$8\frac{3}{16}$	$15\frac{11}{16}$
21	$8\frac{5}{8}$	$16\frac{1}{2}$
22	$9\frac{1}{8}$	$17\frac{1}{4}$
23	$9\frac{7}{16}$	$18\frac{1}{16}$
24	$9\frac{15}{16}$	$18\frac{13}{16}$
25	$10\frac{1}{4}$	$19\frac{9}{16}$
26	$10\frac{11}{16}$	$20\frac{3}{8}$
27	$11\frac{1}{16}$	$21\frac{3}{16}$
28	$11\frac{1}{2}$	22
29	$11\frac{7}{8}$	$22\frac{3}{4}$
30	$12\frac{7}{16}$	$23\frac{9}{16}$
31	$12\frac{13}{16}$	$24\frac{3}{8}$
32	$13\frac{1}{4}$	$25\frac{1}{8}$
33	$13\frac{11}{16}$	$25\frac{15}{16}$
34	$14\frac{1}{16}$	$26\frac{3}{4}$
35	$14\frac{1}{2}$	$27\frac{1}{2}$
36	$14\frac{15}{16}$	$28\frac{1}{4}$
37	$15\frac{5}{16}$	$29\frac{1}{16}$
38	$15\frac{3}{4}$	$29\frac{7}{8}$
39	$16\frac{1}{8}$	$30\frac{5}{8}$
40	$16\frac{9}{16}$	$31\frac{7}{16}$
41	$16\frac{7}{8}$	$32\frac{3}{16}$

SETBACK AND LENGTH OF 45° BENDS WITH RADII FROM 1 TO 72 INCHES (cont.)

Radius of Bend (in.)	Setback (in.)	Length of Bend (in.)
42	17⅜	33
43	17¹³⁄₁₆	33¾
44	18¼	34⁹⁄₁₆
45	18⅝	35⅜
46	19¹⁄₁₆	36⅛
47	19½	36¹⁵⁄₁₆
48	19⅞	37¾
49	20⁵⁄₁₆	38½
50	20¾	39¼
51	21⅛	40¹⁄₁₆
52	21⁹⁄₁₆	40⅞
53	22	41⅝
54	22⅜	42⁷⁄₁₆
55	22¾	43³⁄₁₆
56	23³⁄₁₆	44¹⁄₁₆
57	23⅝	44¾
58	24	45⁹⁄₁₆
59	24⁷⁄₁₆	46⁵⁄₁₆
60	24⅞	47¼
61	25¼	48
62	25¹¹⁄₁₆	48¹³⁄₁₆
63	26⅛	49⅝
64	26⁹⁄₁₆	50⅜
65	26¹⁵⁄₁₆	51³⁄₁₆
66	27⅜	52
67	27¹⁵⁄₁₆	52¾
68	28³⁄₁₆	53½
69	28⅝	54⅜
70	29	55¹⁄₁₆
71	29⁷⁄₁₆	55¹⁵⁄₁₆
72	29¹³⁄₁₆	56¹¹⁄₁₆

SETBACK AND LENGTH OF 60° BENDS

The setback and length of bend for 60° pipe bends are calculated by using these formulas:

Setback = Radius X .577

Length of Bend = Radius X 1.05

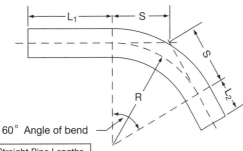

60° Angle of bend

L = Straight Pipe Lengths
R = Bend Radius Length
S = Setback Length

SETBACK AND LENGTH OF 60° BENDS WITH RADII FROM 1 TO 72 INCHES

Radius of Bend (in.)	Setback (in.)	Length of Bend (in.)
1	$9/16$	$1 1/16$
2	$1 1/8$	$2 1/8$
3	$1 3/4$	$3 1/8$
4	$2 5/16$	$4 3/16$
5	$2 7/8$	$5 1/4$
6	$3 1/2$	$6 5/16$
7	$4 1/16$	$7 5/16$
8	$4 5/8$	$8 1/4$
9	$5 3/16$	$9 7/16$
10	$5 13/16$	$10 7/16$

SETBACK AND LENGTH OF 60° BENDS WITH RADII FROM 1 TO 72 INCHES (*cont.*)

Radius of Bend (in.)	Setback (in.)	Length of Bend (in.)
11	6⅜	11½
12	6¹⁵⁄₁₆	12⁹⁄₁₆
13	7½	13⅝
14	8⅛	14¹¹⁄₁₆
15	8¹¹⁄₁₆	15¹¹⁄₁₆
16	9¼	16¾
17	9⅞	17¹³⁄₁₆
18	10⁷⁄₁₆	18¹³⁄₁₆
19	11	19¹⁵⁄₁₆
20	11⁹⁄₁₆	20¹⁵⁄₁₆
21	12³⁄₁₆	22
22	12¾	23¹⁄₁₆
23	13⁵⁄₁₆	24¹⁄₁₆
24	13¹⁵⁄₁₆	25⅛
25	14½	26³⁄₁₆
26	15¹⁄₁₆	27³⁄₁₆
27	15⅝	28⁵⁄₁₆
28	16¼	29⁵⁄₁₆
29	16¹³⁄₁₆	30⅜
30	17⅜	31⁷⁄₁₆
31	17⅞	32½
32	18½	33½
33	19¹⁄₁₆	34⁹⁄₁₆
34	19¹¹⁄₁₆	35⅝
35	20¼	36¹¹⁄₁₆
36	20¹³⁄₁₆	37¾
37	21⅜	38¾
38	22	39¹³⁄₁₆
39	22½	40⅞
40	23⅛	41¹⁵⁄₁₆
41	23¹¹⁄₁₆	42¹⁵⁄₁₆

SETBACK AND LENGTH OF 60° BENDS WITH RADII FROM 1 TO 72 INCHES (*cont.*)		
Radius of Bend (in.)	**Setback (in.)**	**Length of Bend (in.)**
42	24 1/4	44
43	24 7/8	45
44	25 7/16	46 1/8
45	26	47 1/8
46	26 3/8	48 3/16
47	27 1/8	49 1/4
48	27 3/4	50 1/4
49	28 5/16	51 5/16
50	28 7/8	52 3/8
51	29 1/2	53 7/16
52	30	54 1/2
53	30 5/8	55 1/2
54	31 3/16	56 9/16
55	31 3/4	57 5/8
56	32 3/8	58 11/16
57	32 15/16	59 11/16
58	33 1/2	60 3/4
59	34 1/16	61 13/16
60	34 5/8	63
61	35 1/4	64 1/16
62	35 13/16	65 1/8
63	36 3/8	66 1/8
64	37	67 3/6
65	37 9/16	68 1/4
66	38 1/8	69 5/16
67	38 11/16	70 3/8
68	39 1/4	71 7/16
69	39 7/8	72 1/2
70	40 3/8	73 1/2
71	41	74 9/16
72	41 5/8	75 5/8

SETBACK AND LENGTH OF 90° BENDS

The setback and length of bend for 90° pipe bends are calculated by using these formulas:

Setback = Radius X 1.0

Length of Bend = Radius X 1.57

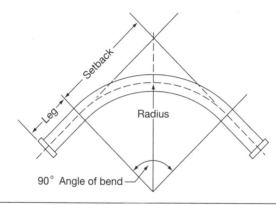

SETBACK AND LENGTH OF 90° BENDS WITH RADII FROM 1 TO 72 INCHES

Radius of Bend (in.)	Setback (in.)	Length of Bend (in.)
1	1	1⁹⁄₁₆
2	2	3⅛
3	3	4¹¹⁄₁₆
4	4	6⁵⁄₁₆
5	5	7⅞
6	6	9⁷⁄₁₆
7	7	11
8	8	12⁹⁄₁₆
9	9	14⅛
10	10	15¹¹⁄₁₆

SETBACK AND LENGTH OF 90° BENDS WITH RADII FROM 1 TO 72 INCHES (cont.)

Radius of Bend (in.)	Setback (in.)	Length of Bend (in.)
11	11	17¼
12	12	18⅞
13	13	20⁷⁄₁₆
14	14	22
15	15	23⁹⁄₁₆
16	16	25⅛
17	17	26¹¹⁄₁₆
18	18	28¼
19	19	29⅞
20	20	31⁷⁄₁₆
21	21	33
22	22	34⁹⁄₁₆
23	23	36⅛
24	24	37¹¹⁄₁₆
25	25	39¼
26	26	40¹³⁄₁₆
27	27	42⁷⁄₁₆
28	28	44
29	29	45⁹⁄₁₆
30	30	47⅛
31	31	48¹¹⁄₁₆
32	32	50¼
33	33	51¹³⁄₁₆
34	34	53⁷⁄₁₆
35	35	55
36	36	56⁹⁄₁₆
37	37	58⅛
38	38	59¾
39	39	61¼
40	40	62⅞
41	41	64⅜

SETBACK AND LENGTH OF 90° BENDS WITH RADII FROM 1 TO 72 INCHES (*cont.*)

Radius of Bend (in.)	Setback (in.)	Length of Bend (in.)
42	42	66
43	43	$67\frac{9}{16}$
44	44	$69\frac{1}{8}$
45	45	$70\frac{11}{16}$
46	46	$72\frac{1}{4}$
47	47	$73\frac{7}{8}$
48	48	$75\frac{7}{16}$
49	49	77
50	50	$78\frac{9}{16}$
51	51	$80\frac{1}{8}$
52	52	$81\frac{11}{16}$
53	53	$83\frac{1}{4}$
54	54	$84\frac{7}{8}$
55	55	$86\frac{3}{8}$
56	56	88
57	57	$89\frac{1}{2}$
58	58	$91\frac{1}{8}$
59	59	$92\frac{11}{16}$
60	60	$94\frac{1}{4}$
61	61	$95\frac{13}{16}$
62	62	$97\frac{1}{2}$
63	63	99
64	64	$100\frac{1}{2}$
65	65	102
66	66	$103\frac{5}{8}$
67	67	$105\frac{7}{16}$
68	68	$106\frac{15}{16}$
69	69	$108\frac{7}{16}$
70	70	110
71	71	$111\frac{1}{2}$
72	72	$113\frac{5}{16}$

22½° OFFSET BENDS

This diagram represents a typical offset bend of a pipe with a 22½° bending angle.

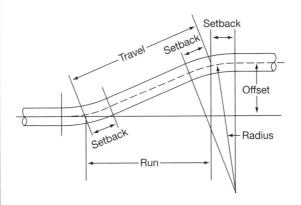

The measurements above may be calculated by using the 22½° offset bend formulas listed below:

$$\text{Offset} = \text{Run X .4142}$$

$$\text{Run} = \text{Offset X 2.4142}$$

$$\text{Travel} = \text{Offset X 2.6131}$$

$$\text{Setback} = \text{Radius X .199}$$

$$\text{Length of Bend} = \text{Radius X .393}$$

22½° OFFSET BENDS FOR
TWO OR MORE PIPES WITH EQUAL SPREAD

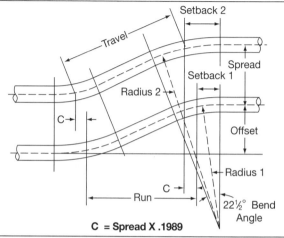

C = Spread X .1989

22½° OFFSET TRAVEL AND RUN DIMENSIONS		
Offset (in.)	Travel (in.)	Run (in.)
¼	⅝	⅝
½	1⁵⁄₁₆	1³⁄₁₆
¾	1¹⁵⁄₁₆	1¹³⁄₁₆
1	2⅝	2⁷⁄₁₆
2	5¼	4¹³⁄₁₆
3	7⅞	7¼
4	10⁷⁄₁₆	9⅝
5	13¹⁄₁₆	12¹⁄₁₆
6	15¹¹⁄₁₆	14¼
7	18⅝	16⅞
8	20⅞	19⁵⁄₁₆
9	23½	21¹¹⁄₁₆
10	26⅛	24⅛
11	28¾	26⁷⁄₁₆
12	31⅜	28⅞

30° OFFSET BENDS

This diagram represents a typical offset bend of a pipe with a 30° bending angle.

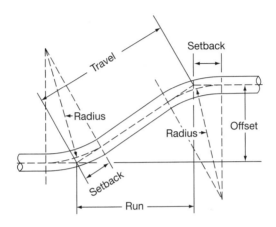

The measurements above may be calculated by using the 30° offset bend formulas listed below:

Offset = Run X .5774

Run = Offset X 1.7321

Travel = Offset X 2.00

Setback = Radius X .268

Length of Bend = Radius X .524

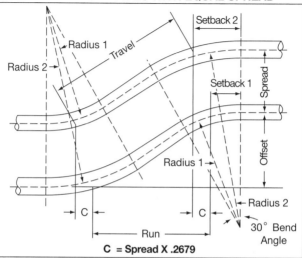

30° OFFSET BENDS FOR TWO OR MORE PIPES WITH EQUAL SPREAD

C = Spread X .2679

30° OFFSET TRAVEL AND RUN DIMENSIONS

Offset (in.)	Travel (in.)	Run (in.)
$\frac{1}{4}$	$\frac{1}{2}$	$\frac{7}{16}$
$\frac{1}{2}$	1	$\frac{7}{8}$
$\frac{3}{4}$	$1\frac{1}{2}$	$1\frac{5}{16}$
1	2	$1\frac{3}{4}$
2	4	$3\frac{5}{16}$
3	6	$5\frac{3}{16}$
4	8	$6\frac{15}{16}$
5	10	$8\frac{5}{8}$
6	12	$10\frac{1}{2}$
7	14	$12\frac{1}{8}$
8	16	$13\frac{7}{8}$
9	18	$15\frac{9}{16}$
10	20	$17\frac{5}{16}$
11	22	19
12	24	$20\frac{13}{16}$

2-21

45° OFFSET BENDS

This diagram represents a typical offset bend of a pipe with a 45° bending angle.

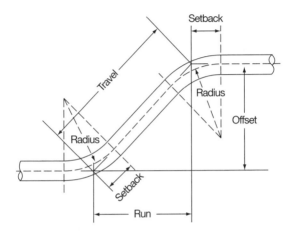

The measurements above may be calculated by using the 45° offset bend formulas listed below:

Offset = Run X 1.00

Run = Offset X 1.00

Travel = Offset X 1.4142

Setback = Radius x .410

Length of Bend = Radius X .780

45° OFFSET BENDS FOR
TWO OR MORE PIPES WITH EQUAL SPREAD

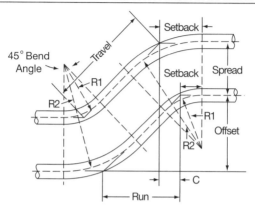

C = Spread X .4142

45° OFFSET TRAVEL AND RUN DIMENSIONS

Offset (in.)	Travel (in.)	Run (in.)
¼	⅜	¼
½	¾	½
¾	1¹⁄₁₆	¾
1	1⁷⁄₁₆	1
2	2¹³⁄₁₆	2
3	4¼	3
4	5⅝	4
5	7¹⁄₁₆	5
6	8½	6
7	9⅞	7
8	11⁵⁄₁₆	8
9	12¾	9
10	14⅛	10
11	15⁹⁄₁₆	11
12	17	12

60° OFFSET BENDS

This diagram represents a typical offset bend of a pipe with a 60° bending angle.

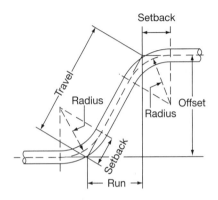

The measurements above may be calculated by using the 60° offset bend formulas listed below:

$$\text{Offset} = \text{Run X 1.732}$$

$$\text{Run} = \text{Offset X .5774}$$

$$\text{Travel} = \text{Offset X 1.1547}$$

$$\text{Setback} = \text{Radius X .580}$$

$$\text{Length of Bend} = \text{Radius X 1.05}$$

60° OFFSET BENDS FOR
TWO OR MORE PIPES WITH EQUAL SPREAD

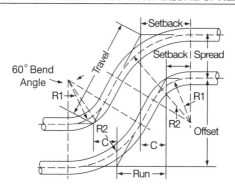

$$C = Spread \times .5774$$

60° OFFSET TRAVEL AND RUN DIMENSIONS

Offset (in.)	Travel (in.)	Run (in.)
1/4	5/16	1/8
1/2	9/16	5/16
3/4	7/8	7/16
1	1 1/8	9/16
2	2 5/16	1 1/8
3	3 7/16	1 3/4
4	4 5/8	2 5/16
5	5 3/4	2 7/8
6	6 15/16	3 7/16
7	8 1/16	4 1/16
8	9 1/4	4 5/8
9	10 3/8	5 3/16
10	11 9/16	5 3/4
11	12 11/16	6 3/8
12	13 7/8	6 15/16

90° OFFSET BENDS FOR
TWO OR MORE PIPES WITH EQUAL SPREAD

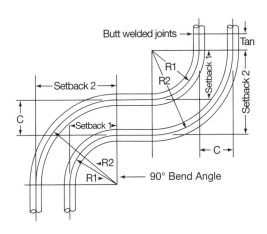

Offset = any dimension greater than Radius X 2.0

Run = 0

Travel = Offset

Setback 1 = Radius 1

Setback 2 = Radius 2

Length of Bend = Radius X 1.57

C = Spread

CONTINUOUS BEND OFFSETS

Where offsets are created by two equal and opposite bends with no straight pipe between the bends, the offset, end-to-end run and length of pipe dimensions are fixed by the radius and bending angle.

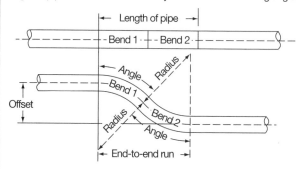

Offset Bend With Two 30° Bends				Offset Bend With Two 45° Bends			
Radius (in.)	Offset (in.)	E to E Run (in.)	Length of Pipe (in.)	Radius (in.)	Offset (in.)	E to E Run (in.)	Length of Pipe (in.)
2	½	2	2⅛	2	1³⁄₁₆	2¹³⁄₁₆	3⅛
4	1	4	4³⁄₁₆	4	2⅜	5⅝	6⁵⁄₁₆
6	1½	6	6⁵⁄₁₆	6	3½	8½	9⁷⁄₁₆
8	2	8	8¼	8	4¹¹⁄₁₆	11⁵⁄₁₆	12⁹⁄₁₆
10	2½	10	10⁷⁄₁₆	10	5⅞	14⅛	15¹¹⁄₁₆
12	3	12	12⁹⁄₁₆	12	7	17	18⅞

For all angles and radii in continuous bend offsets, the formulas are as follows:

Offset = 2.0 X Radius X (1 – Sine of Angle)

End-to-End Run = 2.0 X Radius X Sine of Angle

Length of Pipe = .035 X Radius X No. of Degrees in Angle

180° CROSSOVER BEND

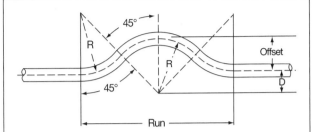

The 180° crossover bend is made up of 4 continuous 45° bends. When the radius dimension is known, the following calculations can be used.

$$\text{Offset} = \text{Radius} \times .5858$$
$$\text{Run} = \text{Radius} \times 2.828$$
$$\text{D} = \text{Radius} \times .4142$$
$$\text{D} = \text{Radius} - \text{Offset}$$
$$\text{Length of Pipe in Bend} = \text{Radius} \times 3.1416$$

If offset is known, then Radius = Offset X 1.707

CALCULATED DIMENSIONS

Radius (in.)	Offset (in.)	Run (in.)	Length of Pipe in Bend (in.)
4	2⅜	11⅜	12⁹⁄₁₆
6	3½	17	18⅞
8	4¹¹⁄₁₆	22⁹⁄₁₆	25⅛
10	5⅞	28¼	31⁷⁄₁₆
12	7	33⅞	37¾
14	8³⁄₁₆	39½	44
16	9⅜	45⁷⁄₁₆	50¼
18	10½	50⅞	56⅝

ADDITIONAL 180° BENDS

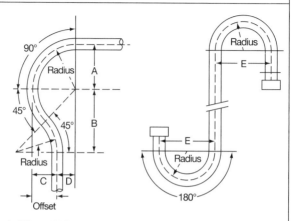

Single Offset 180° Quarter Bend

Two 180° U-bends

The 180° quarter bend is made up of two 45° bends and one 90° bend. The U-bends are continuous arcs of 180° each. Despite what the direction of bending is, if the radius is constant, then the length of pipe in a 180° bending total is equal to the radius multiplied by 3.1416.

Length of Pipe in 180° Bends = Radius X 3.1416

When offset is known, then Radius = Offset C X 1.707

The drawing dimensions are calculated as follows:

> **Dimension A = Radius**
> **Dimension B = Radius X 1.4142**
> **Dimension C = Radius X .5858**
> **Dimension D = Radius X .4142**
> **Dimension E = Radius X 2.0**

210° SINGLE OFFSET QUARTER BEND

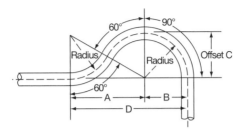

Dimension A = Radius X 1.732

Dimension B = Radius

Dimension C = Radius = Offset

Dimension D = Radius X 2.732

Length of Bend = Radius X 3.6651

CALCULATED DIMENSIONS

Radius (in.)	Length of Bend (in.)	D (in.)	Radius (in.)	Length of Bend (in.)	D (in.)
1	3⅝	2¾	18	66	49¼
2	7⅜	5½	24	88	65½
3	11	8¼	36	132	98¼
4	14¾	11	48	176	131¼
5	18¼	13¾	60	220	164
6	22	16½	72	264	196½
7	25¾	19¼	84	308	229½
8	29¼	22	96	351¾	262
9	33	24½	108	395¾	295
10	36¾	27¼	120	439¾	328
11	40¼	30	132	483¾	360½
12	44	32¾	144	527¾	393½

240° CROSSOVER BEND

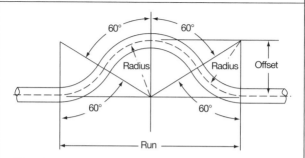

When the radius is known, the dimensions below are calculated as follows:

$$\text{Offset} = \text{Radius}$$

$$\text{Run} = \text{Radius} \times 3.464$$

$$\text{Length of Bend} = \text{Radius} \times 4.1887$$

CALCULATED DIMENSIONS

Radius (in.)	Run (in.)	Length of Bend (in.)
4	13⅞	16¾
6	20¾	25⅛
8	27¾	33½
10	34⅝	41⅞
12	41⅝	50¼
14	48½	58⅝
16	55½	67
18	62⅜	75½
20	69¼	83¾
22	76¼	92⅛
24	83⅛	100½

270° SINGLE OFFSET U-BEND

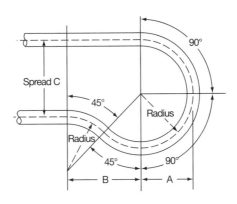

Dimension A = Bend Radius

Dimension B or C = Radius X 1.4142

Length of Bend = Radius X 4.7123

If spread is known, then Radius = Spread C X .707

CALCULATED DIMENSIONS

Radius (in.)	Length of Bend (in.)	B or C (in.)	Radius (in.)	Length of Bend (in.)	B or C (in.)
2	9½	2¾	12	56½	16⅞
3	14¼	4¼	18	84¾	25½
4	18¾	5⅝	24	113	34
5	23½	7⅛	36	169½	50¾
6	28¼	8½	48	226¼	68
7	33	9⅞	60	282¾	85
8	37¾	11¼	72	339¼	100¼
9	42½	12¾	84	395¾	118½
10	47¼	14⅛	96	452½	135¾
11	51¾	15½	120	565½	170

300° SINGLE OFFSET U-BEND

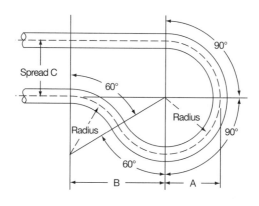

Dimension A = Bend Radius

Dimension B = Radius X 1.732

Spread C = Radius

Length of Bend = Radius X 5.2359

CALCULATED DIMENSIONS

Radius (in.)	Length of Bend (in.)	B (in.)	Radius (in.)	Length of Bend (in.)	B (in.)
2	10½	3½	12	62¾	20¾
3	15¾	5¼	18	94¼	31¼
4	21	6¾	24	125¾	41½
5	26¼	8¾	36	188½	62½
6	31½	10½	48	251¼	83¼
7	36¾	12¼	60	314¼	103¾
8	42	13¾	72	377	124½
9	47¼	15½	84	439¾	141¼
10	52½	17¼	96	502¾	166
11	57½	19	120	754	207½

360° DOUBLE OFFSET AND EXPANSION U-BENDS

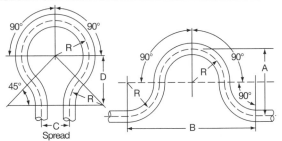

Double offset U-bend
(R=Radius)

Expansion U-bend
(R=Radius)

Dimension A = Radius X 2
Dimension B = Radius X 4
Dimension D = Radius X 1.4142
Spread C = Radius X .8284
Radius = Spread C X 1.207
Length of Bend = Radius X 6.2832

CALCULATED DIMENSIONS

Radius (in.)	Length of Bend (in.)	C (in.)	Radius (in.)	Length of Bend (in.)	C (in.)
2	12½	1⅝	18	113⅛	10¾
3	18⅞	2½	24	150⅞	19¾
4	25⅛	3¼	36	226¼	29
5	31⅜	4⅛	48	301¾	39¾
6	37⅝	5	60	377⅛	49¾
7	44	5¾	72	452½	59⅝
8	50¼	6⅝	84	527¾	69½
9	56½	7½	96	603	79½
10	62⅞	8¼	120	754	99⅜
11	69	9⅛			
12	75⅜	9⅞			

540° EXPANSION U-BEND

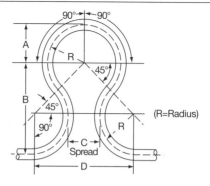

Dimension A = Radius
Dimension B = Radius X 2.4142
Dimension D = Radius X 2.8284
Spread C = Radius X .8284
Radius = Spread C X 1.207
Length of Bend = Radius X 9.4246

CALCULATED DIMENSIONS

Radius (in.)	Length of Bend (in.)	B (in.)	C (in.)	D (in.)
6	56⅝	14½	5	17
7	66	16⅞	5¾	19¾
8	75⅜	19¼	6⅝	22⅝
9	84¾	21¾	7½	25½
10	94¼	24⅛	8¼	28¼
12	113⅛	28⅞	10	34
18	169¾	43⅜	14⅞	50⅞
24	226⅛	57⅞	19⅞	62⅞
36	339¼	81¾	29¾	101⅞
48	452½	115⅞	39¾	135¾

2-35

BENDING PIPE AROUND A CIRCLE

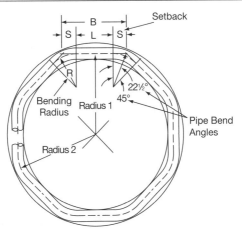

**The bending radius (R) may be any length shorter than
Radius 1 Dimension L = B − [2 X Setback]**

		To Find Dimension B	
Number of Sides of Pipe	**Pipe Bend Angles in Degrees**	**For Pipe Within Circle Multiply Radius 2 by**	**For Pipe Outside Circle Multiply Radius 1 by**
4	90°	1.414	2.000
5	72°	1.175	1.453
6	60°	1.000	1.155
8	45°	.765	.828
9	40°	.685	.728
10	36°	.618	.650
12	30°	.518	.536
16	22½°	.390	.398
20	18°	.313	.317
24	15°	.261	.263

MINIMUM BENDING RADIUS FOR STEEL AND WROUGHT IRON PIPE*

Nominal Pipe Size (in.)	Minimum Bending Radius (in.)	
	Steel Pipe	Wrought Iron Pipe
1/4	1	—
3/8	1 1/4	—
1/2	1 1/2	1 3/8
3/4	1 3/4	1 3/4
1	2	2 1/8
1 1/4	2 1/4	2 3/4
1 1/2	2 1/2	3 1/2
2	3	5 1/2
2 1/2	5	10
3	8	12
3 1/2	10	14
4	12	16
5	18	20
6	22	26
8	30	30
10	36	36
12	46	46

*Applies to cold bending where some flattening at bend is acceptable.

BASIC OFFSET CONNECTIONS

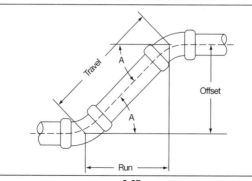

BASIC OFFSET CONNECTIONS FOR TWO OR MORE PIPES WITH EQUAL SPREAD

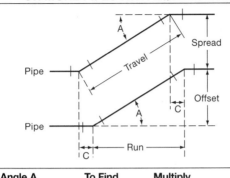

Angle A	To Find	Multiply	by
22½°	Offset	Run	.414
	Run	Offset	2.414
	Travel	Offset	2.613
	C	Spread	.199
30°	Offset	Run	.577
	Run	Offset	1.732
	Travel	Offset	2.000
	C	Spread	.268
45°	Offset	Run	1.000
	Run	Offset	1.000
	Travel	Offset	1.414
	C	Spread	.414
60°	Offset	Run	1.732
	Run	Offset	.577
	Travel	Offset	1.155
	C	Spread	.577
Any Angle	Offset	Run	Tan A
	Offset	Travel	Sin A
	Run	Offset	Cot A
(See Trig. Tables on Page 2-58)	Travel	Offset	Csc A
	C	Spread	Tan ½ A

ROLLING OFFSET

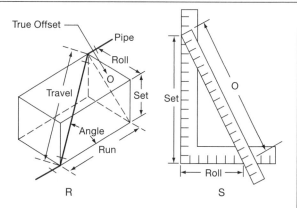

In the rolling offset depicted in drawing R, the plane of the travel determines that the diagonal O is the true offset. The following are two methods of finding the travel and run:

(1) Manual method: Mark off the roll and set on a steel square as shown in drawing S.

The straight line distance is dimension O which is to be read from the tables on preceding pages in this chapter.

(2) Mathematical method:

The same basic formula can be applied to any offset problem by interchanging the terms as follows:

$$O = \sqrt{(Roll)^2 + (Set)^2} \text{ and } O^2 = (Roll)^2 + (Set)^2$$

The same basic formula can be applied to any offset problem by changing the terms as follows:

$$Travel = \sqrt{(Offset)^2 + (Run)^2}$$

Thus, for the offset above, $Travel = \sqrt{(O)^2 + (Run)^2}$

VARIOUS TWO-PIPE 90° TURNS WITH EQUAL SPREAD

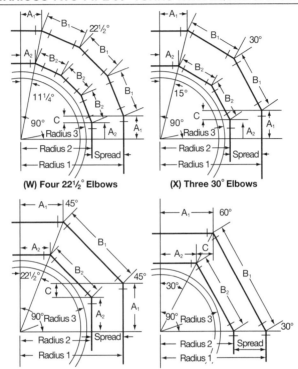

(W) Four 22½° Elbows

(X) Three 30° Elbows

(Y) Two 45° Elbows

(Z) One 30° and One 60° Elbow

To Find	Multiply	For W by	For X by	For Y by	For Z by
A_1	Radius 1	.199	.268	.414	.577
A_2	Radius 2	.199	.268	.414	.577
B_1	Radius 1	.398	.536	.828	1.155
B_2	Radius 2	.398	.536	.828	1.155
C	Spread	.199	.268	.414	.577

SCREWED PIPE COILS INSIDE OR OUTSIDE OF ROUND TANKS

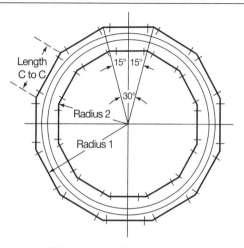

In this example 30° elbows are used.

Fitting Angle in Degrees	Number of Elbows or Straight Pieces Per Coil	To Find C to C Dimension	
		If Inside Coil Multiply Radius 2 by	If Outside Coil Multiply Radius 1 by
22½°	16	.390	.398
30°	12	.518	.536
45°	8	.765	.828
60°	6	1.000	1.155
90°	4	1.414	2.000
*A°	360/A	2 X Sin ½ A	2 X Tan ½ A

*A equals any angle evenly divisible by 360.

MITERING PIPE

A miter cut is made at half the angle of the completed joint. In fabricating welded offset connections, the cut is positioned as shown in the drawing below. For the highest accuracy, use the actual O.D. of the pipe as given in manufacturer's literature, not the NPS (I.D.).

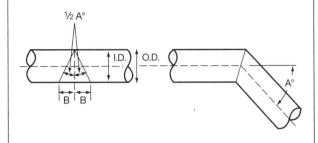

For a Mitered Offset When Angle A° is Equal to	Distance B is Equal to the Actual Pipe O.D. Times	Nominal Pipe Size (I.D.)			
		2"	3"	4"	6"
		Distance B (in.)			
22½°	.199	.472	.696	.895	1.32
30°	.268	.635	.937	1.208	1.97
45°	.414	.982	1.45	1.86	2.76
60°	.577	1.37	2.02	2.60	3.82
90°	1.000	2.37	3.50	4.50	6.63
Any Angle	Tan ½ A	Values Above Nearest 1/16 In.			

The tangents of all angles are given in the trigonometry section which begins on page 2-58.

3-PIECE 45° WELDED TURN

In a 3-piece 45° turn, the mitered pipe is always cut at 11¼° to make two 22½° joints. The angle of the miter cut always has its apex at the center from which the radius of the turn is drawn. Thus,

There are two mitered joints in a 3-piece 45° turn
Each joint is 45° ÷ 2 = 22½°
Each piece is mitered at 22½° ÷ 2 = 11¼°

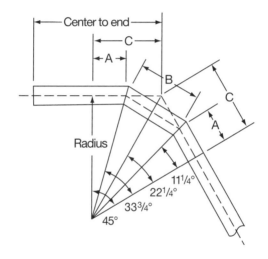

Dimension calculations for the drawing above are:

Dimension A = Radius X .1989
Dimension B = Radius X .3978
Dimension B = A X 2.0
Dimension C = Radius X .4142

3-PIECE 45° WELDED TURN DIMENSIONS

Radius of Turn (in.)	A (in.)	B (in.)	C (in.)
1	$3/16$	$3/8$	$7/16$
2	$3/8$	$13/16$	$13/16$
3	$5/8$	$1\,3/16$	$1\,1/4$
4	$13/16$	$1\,5/8$	$1\,5/8$
5	1	2	$2\,1/16$
6	$1\,3/16$	$2\,3/8$	$2\,7/16$
7	$1\,3/8$	$2\,3/4$	$2\,7/8$
8	$1\,5/8$	$3\,3/16$	$3\,1/4$
9	$1\,13/16$	$3\,9/16$	$3\,3/4$
10	2	4	$4\,1/8$
11	$2\,3/16$	$4\,3/8$	$4\,1/2$
12	$2\,3/8$	$4\,3/4$	$4\,15/16$
13	$2\,9/16$	$5\,3/16$	$5\,5/16$
14	$2\,3/4$	$5\,9/16$	$5\,3/4$
15	3	6	$6\,1/8$
16	$3\,3/16$	$6\,3/8$	$6\,9/16$
17	$3\,3/8$	$6\,3/4$	7
18	$3\,9/16$	$7\,3/16$	$7\,7/16$
19	$3\,3/4$	$7\,9/16$	$7\,7/8$
20	4	$7\,15/16$	$8\,3/16$
21	$4\,3/16$	$8\,3/8$	$8\,5/8$
22	$4\,3/8$	$8\,3/4$	$9\,1/8$
23	$4\,9/16$	$9\,1/8$	$9\,7/16$
24	$4\,3/4$	$9\,9/16$	$9\,11/16$

3-PIECE 45° WELDED TURN DIMENSIONS (*cont.*)

Radius of Turn (in.)	A (in.)	B (in.)	C (in.)
25	5	9$\frac{15}{16}$	10$\frac{1}{4}$
26	5$\frac{3}{16}$	10$\frac{3}{8}$	10$\frac{11}{16}$
27	5$\frac{3}{8}$	10$\frac{3}{4}$	11$\frac{1}{16}$
28	5$\frac{9}{16}$	11$\frac{1}{8}$	11$\frac{1}{2}$
29	5$\frac{3}{4}$	11$\frac{1}{2}$	11$\frac{7}{8}$
30	6	11$\frac{15}{16}$	12$\frac{7}{16}$
31	6$\frac{3}{16}$	12$\frac{5}{16}$	12$\frac{13}{16}$
32	6$\frac{3}{8}$	12$\frac{3}{4}$	13$\frac{1}{4}$
33	6$\frac{9}{16}$	13$\frac{1}{8}$	13$\frac{11}{16}$
34	6$\frac{3}{4}$	13$\frac{1}{2}$	14$\frac{1}{16}$
35	6$\frac{15}{16}$	13$\frac{15}{16}$	14$\frac{1}{2}$
36	7$\frac{3}{16}$	14$\frac{5}{16}$	14$\frac{15}{16}$
37	7$\frac{3}{8}$	14$\frac{11}{16}$	15$\frac{5}{16}$
38	7$\frac{9}{16}$	15$\frac{1}{8}$	15$\frac{3}{4}$
39	7$\frac{3}{4}$	15$\frac{1}{2}$	16$\frac{1}{8}$
40	7$\frac{15}{16}$	15$\frac{15}{16}$	16$\frac{9}{16}$
41	8$\frac{3}{16}$	16$\frac{5}{16}$	17
42	8$\frac{3}{8}$	16$\frac{11}{16}$	17$\frac{3}{8}$
43	8$\frac{9}{16}$	17$\frac{1}{8}$	17$\frac{13}{16}$
44	8$\frac{3}{4}$	17$\frac{1}{2}$	18$\frac{3}{16}$
45	8$\frac{15}{16}$	17$\frac{15}{16}$	18$\frac{5}{8}$
46	9$\frac{3}{16}$	18$\frac{5}{16}$	19$\frac{1}{16}$
47	9$\frac{3}{8}$	18$\frac{11}{16}$	19$\frac{7}{16}$
48	9$\frac{9}{16}$	19$\frac{1}{8}$	19$\frac{7}{8}$

3-PIECE 60° WELDED TURN

In a 3-piece 60° turn, the mitered pipe is always cut at 15° to make two 30° joints. The angle of the miter cut always has its apex at the center from which the radius of the turn is drawn. Thus,

There are two mitered joints in a 3-piece 60° turn
Each joint is 60° ÷ 2 = 30°
Each piece is mitered at 30° ÷ 2 =15°

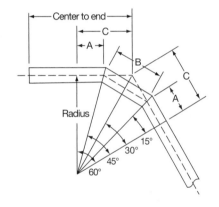

Dimension calculations for the drawing above are:

> **Dimension A = Radius X .2679**
> **Dimension B = Radius X .5358**
> **Dimension B = A X 2.0**
> **Dimension C = Radius X .5774**

3-PIECE 60° WELDED TURN DIMENSIONS

Radius of Turn (in.)	A (in.)	B (in.)	C (in.)
1	¼	$9/16$	$9/16$
2	$9/16$	$1\,1/16$	$1\,1/8$
3	$13/16$	$1\,5/8$	$1\,3/4$
4	$1\,1/16$	$2\,1/8$	$2\,5/16$
5	$1\,5/16$	$2\,11/16$	$2\,7/8$
6	$1\,5/8$	$3\,3/16$	$3\,1/2$
7	$1\,7/8$	$3\,3/4$	$4\,1/16$
8	$2\,1/8$	$4\,5/16$	$4\,5/8$
9	$2\,7/16$	$4\,7/8$	$5\,3/16$
10	$2\,11/16$	$5\,3/8$	$5\,13/16$
11	$2\,15/16$	$5\,7/8$	$6\,3/8$
12	$3\,3/16$	$6\,7/16$	$6\,15/16$
13	$3\,1/2$	$6\,15/16$	$7\,1/2$
14	$3\,3/4$	$7\,1/2$	$8\,1/8$
15	4	8	$8\,11/16$
16	$4\,5/16$	$8\,9/16$	$9\,1/4$
17	$4\,9/16$	$9\,1/8$	$9\,7/8$
18	$4\,13/16$	$9\,5/8$	$10\,7/16$
19	$5\,1/16$	$10\,3/16$	11
20	$5\,3/8$	$10\,3/4$	$11\,9/16$
21	$5\,5/8$	$11\,1/4$	$12\,3/16$
22	$5\,7/8$	$11\,13/16$	$12\,3/4$
23	$6\,3/16$	$12\,5/16$	$13\,5/16$
24	$6\,7/16$	$12\,7/8$	$13\,15/16$

3-PIECE 60° WELDED TURN DIMENSIONS (*cont.*)

Radius of Turn (in.)	A (in.)	B (in.)	C (in.)
25	6¹¹⁄₁₆	13⅜	14½
26	7	13¹⁵⁄₁₆	15¹⁄₁₆
27	7¼	14½	15⅝
28	7½	15	16¼
29	7¾	15½	16¹³⁄₁₆
30	8¹⁄₁₆	16¹⁄₁₆	17⅜
31	8⁵⁄₁₆	16⅝	17⅞
32	8⁹⁄₁₆	17⅛	18½
33	8⅞	17¹¹⁄₁₆	19¹⁄₁₆
34	9⅛	18³⁄₁₆	19¹¹⁄₁₆
35	9⅜	18¾	20¼
36	9⅝	19⁵⁄₁₆	20¹³⁄₁₆
37	9¹⁵⁄₁₆	19⅞	21⅜
38	10³⁄₁₆	20⅜	21⅞
39	10⁷⁄₁₆	20¹⁵⁄₁₆	22½
40	10¹¹⁄₁₆	21⁷⁄₁₆	23⅛
41	11	22	23¹¹⁄₁₆
42	11¼	22⁹⁄₁₆	24¼
43	11½	23¹⁄₁₆	24¹³⁄₁₆
44	11¾	23⅝	25⁷⁄₁₆
45	12¹⁄₁₆	24⅛	26
46	12⁵⁄₁₆	24¹¹⁄₁₆	26⁹⁄₁₆
47	12⁹⁄₁₆	25³⁄₁₆	27⅛
48	12⅞	25¾	27¹¹⁄₁₆

3-PIECE 90° WELDED TURN

In a 3-piece 90° turn, the mitered pipe is always cut at 22½° to make two 45° joints. Even though the drawing shows a 30° angle with sides B and C, it is only used to determine those sides and does not set the angle of the miter cut. Thus,

There are two mitered joints in a 3-piece 90° turn
Each joint is 90° ÷ 2 = 45°
Each piece is mitered at 45° ÷ 2 = 22½°

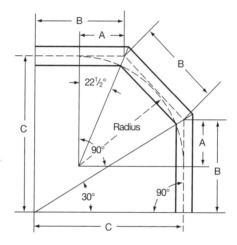

Dimension calculations for the drawing above are:

Dimension A = Radius X .4142

Dimension B = Radius X .828

Dimension B = A X 2.0

Dimension B = C X .5774

Dimension C = B X 1.732

2-49

3-PIECE 90° WELDED TURN DIMENSIONS

Radius of Turn (in.)	A (in.)	B (in.)
1	$7/16$	$13/16$
2	$13/16$	$1 \, 11/16$
3	$1 \, 1/4$	$2 \, 1/2$
4	$1 \, 11/16$	$3 \, 5/16$
5	$2 \, 1/16$	$4 \, 1/8$
6	$2 \, 1/2$	5
7	$2 \, 7/8$	$5 \, 13/16$
8	$3 \, 5/16$	$6 \, 5/8$
9	$3 \, 3/4$	$7 \, 7/16$
10	$4 \, 1/8$	$8 \, 5/16$
11	$4 \, 9/16$	$9 \, 1/8$
12	5	$9 \, 15/16$
13	$5 \, 3/8$	$10 \, 3/4$
14	$5 \, 13/16$	$11 \, 5/8$
15	$6 \, 3/16$	$12 \, 7/16$
16	$6 \, 5/8$	$13 \, 1/4$
17	$7 \, 1/16$	$14 \, 1/16$
18	$7 \, 7/16$	$14 \, 15/16$
19	$7 \, 7/8$	$15 \, 3/4$
20	$8 \, 5/16$	$16 \, 9/16$
21	$8 \, 11/16$	$17 \, 3/8$
22	$9 \, 1/8$	$18 \, 1/4$
23	$9 \, 1/2$	$19 \, 1/16$
24	$9 \, 15/16$	$19 \, 7/8$

3-PIECE 90° WELDED TURN DIMENSIONS (*cont.*)

Radius of Turn (in.)	A (in.)	B (in.)
25	$10\frac{3}{8}$	$20\frac{11}{16}$
26	$10\frac{3}{4}$	$21\frac{9}{16}$
27	$11\frac{3}{16}$	$22\frac{3}{8}$
28	$11\frac{5}{8}$	$23\frac{3}{16}$
29	12	24
30	$12\frac{7}{16}$	$24\frac{7}{8}$
31	$12\frac{13}{16}$	$25\frac{11}{16}$
32	$13\frac{1}{4}$	$26\frac{1}{2}$
33	$13\frac{11}{16}$	$27\frac{5}{16}$
34	$14\frac{1}{16}$	$28\frac{3}{16}$
35	$14\frac{1}{2}$	29
36	$14\frac{15}{16}$	$29\frac{13}{16}$
37	$15\frac{5}{16}$	$30\frac{5}{8}$
38	$15\frac{3}{4}$	$31\frac{7}{16}$
39	$16\frac{1}{8}$	$32\frac{5}{16}$
40	$16\frac{9}{16}$	$33\frac{1}{8}$
41	$16\frac{15}{16}$	$33\frac{15}{16}$
42	$17\frac{3}{8}$	$34\frac{13}{16}$
43	$17\frac{13}{16}$	$35\frac{5}{8}$
44	$18\frac{3}{16}$	$36\frac{7}{16}$
45	$18\frac{5}{8}$	$37\frac{1}{4}$
46	19	$38\frac{1}{8}$
47	$19\frac{7}{16}$	$38\frac{7}{8}$
48	$19\frac{7}{8}$	$39\frac{3}{4}$

4-PIECE 90° WELDED TURN

Even though the drawing shows a 22½° angle with sides B and C, it is only used to determine the length of B or C when one or the other side is known and does not set the actual angle of the miter cut. Thus,

There are three mitered joints in a 4-piece 90° turn
Each joint is 90° ÷ 3 = 30°
Each piece is mitered at 30° ÷ 2 = 15°

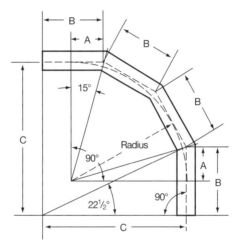

Dimension calculations for the drawing above are:

> **Dimension A = Radius X .2679**
> **Dimension B = Radius X .536**
> **Dimension B = A X 2.0**
> **Dimension B = C X .4142**
> **Dimension C = B X 2.4142**

4-PIECE 90° WELDED TURN DIMENSIONS		
Radius of Turn (in.)	A (in.)	B (in.)
1	¼	9/16
2	9/16	1 1/16
3	13/16	1 5/8
4	1 1/16	2 1/8
5	1 3/8	2 11/16
6	1 5/8	3 3/16
7	1 7/8	3 3/4
8	2 1/8	4 5/16
9	2 7/16	4 7/8
10	2 11/16	5 3/8
11	2 15/16	5 7/8
12	3 1/4	6 7/16
13	3 1/2	6 15/16
14	3 3/4	7 1/2
15	4	8
16	4 5/16	8 9/16
17	4 9/16	9 1/8
18	4 13/16	9 5/8
19	5 1/8	10 3/16
20	5 3/8	10 3/4
21	5 5/8	11 1/4
22	5 7/8	11 13/16
23	6 3/16	12 5/16
24	6 7/16	12 7/8

4-PIECE 90° WELDED TURN DIMENSIONS (*cont.*)

Radius of Turn (in.)	A (in.)	B (in.)
25	$6^{11}/_{16}$	$13^{3}/_{8}$
26	7	$13^{15}/_{16}$
27	$7^{1}/_{4}$	$14^{1}/_{2}$
28	$7^{1}/_{2}$	15
29	$7^{3}/_{4}$	$15^{1}/_{2}$
30	$8^{1}/_{16}$	$16^{1}/_{16}$
31	$8^{5}/_{16}$	$16^{5}/_{8}$
32	$8^{9}/_{16}$	$17^{1}/_{8}$
33	$8^{7}/_{8}$	$17^{11}/_{16}$
34	$9^{1}/_{8}$	$18^{3}/_{16}$
35	$9^{3}/_{8}$	$18^{3}/_{4}$
36	$9^{11}/_{16}$	$19^{5}/_{16}$
37	$9^{15}/_{16}$	$19^{13}/_{16}$
38	$10^{3}/_{16}$	$20^{3}/_{8}$
39	$10^{7}/_{16}$	$20^{15}/_{16}$
40	$10^{3}/_{4}$	$21^{7}/_{16}$
41	11	22
42	$11^{1}/_{4}$	$22^{1}/_{2}$
43	$11^{1}/_{2}$	$23^{1}/_{16}$
44	$11^{13}/_{16}$	$23^{5}/_{8}$
45	$12^{1}/_{16}$	$24^{1}/_{8}$
46	$12^{5}/_{16}$	$24^{11}/_{16}$
47	$12^{5}/_{8}$	$25^{3}/_{16}$
48	$12^{7}/_{8}$	$25^{3}/_{4}$

MITERING PIPE AROUND A CIRCLE

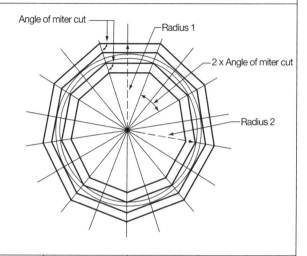

Angle of miter cut

Radius 1

2 x Angle of miter cut

Radius 2

Number of Sides	Miter Cut Angle in Degrees	To Find Length of Straight Side Located on Outside Edge	
		Inside Circle Multiply Radius 2 by	Outside Circle Multiply Radius 1 by
4	45°	1.414	2.000
5	36°	1.175	1.453
6	30°	1.000	1.155
8	22½°	.765	.828
9	20°	.685	.728
10	18°	.618	.650
12	15°	.518	.536
16	11¼°	.390	.398
20	9°	.313	.317
X	$\frac{360}{2X}$	$2 \times \mathrm{Sin}\, \frac{360}{2X}$	$2 \times \mathrm{Tan}\, \frac{360}{2X}$

90° BRACKET CONSTRUCTION WITH DIAGONAL LEG AT ANY ANGLE

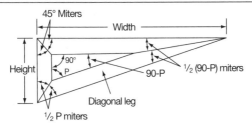

The pitch of the diagonal leg is at angle P from wall, thus

Width = Height X Tan P = Diagonal Leg X Sin P
Height = Width X Cot P = Diagonal Leg X Cos P
Diagonal Leg = Height X Sec P = Width X Csc P

CUTTING PIECES FOR THE BRACKET

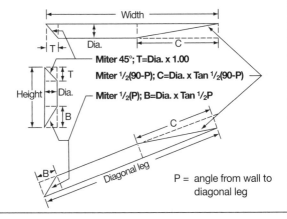

When mitering round pieces, you must be absolutely sure that the cuts on both ends are based on a single centerline.

LENGTH OF EQUAL SEGMENTS FOR THE CIRCUMFERENCE OF VARIOUS SCHEDULE 40 STEEL PIPE SIZES

Nominal Pipe Size (in.)	Actual O. D. (in.)	Actual Circum. (in.)	Total Number of Segments				
			4	6	8	10	12
			Length of each Segment (in.)				
1¼	1.660	5.215	1.30	.87	.65	.52	.43
1½	1.900	5.969	1.49	1.00	.75	.60	.50
2	2.375	7.461	1.87	1.24	.93	.75	.62
2½	2.875	9.032	2.26	1.51	1.13	.90	.75
3	3.500	10.996	2.75	1.67	1.37	1.10	.92
3½	4.000	12.566	3.14	2.10	1.57	1.26	1.05
4	4.500	14.137	3.54	2.36	1.75	1.41	1.18
5	5.563	17.477	4.37	2.92	2.18	1.75	1.46
6	6.625	20.813	5.20	3.47	2.60	2.08	1.73
8	8.625	27.096	6.77	4.52	3.39	2.71	2.26
10	10.750	33.772	8.45	5.60	4.23	3.38	2.82
12	12.750	40.055	10.00	6.67	5.00	4.01	3.34
14	14.000	44.000	11.00	7.35	5.51	4.40	3.66
16	16.000	50.375	12.60	8.40	6.30	5.04	4.19
18	18.000	56.549	14.13	9.41	7.06	5.65	4.70
20	20.000	62.832	15.70	10.48	7.85	6.28	5.22

TRIGONOMETRIC FUNCTIONS

Degrees	Sin	Cos	Tan	Cot	Sec	Csc
0	0	1.00000	0	Inf.	1.0000	Inf.
½	.00873	.99996	.00873	114.59	1.0000	114.59
1	.01745	.99985	.01745	57.290	1.0001	57.299
1½	.02618	.99966	.02618	38.188	1.0003	38.201
2	.03490	.99939	.03492	28.636	1.0006	28.654
2½	.04362	.99905	.04366	22.904	1.0009	22.925
3	.05234	.99863	.05241	19.081	1.0014	19.107
3½	.06105	.99813	.06116	16.350	1.0019	16.380
4	.06976	.99756	.06993	14.301	1.0024	14.335
4½	.07846	.99692	.07870	12.706	1.0031	12.745
5	.08715	.99619	.08749	11.430	1.0038	11.474
5½	.09584	.99540	.09629	10.385	1.0046	10.433
6	.01453	.99452	.10510	9.5144	1.0055	9.5668
6½	.11320	.99357	.11393	8.7769	1.0065	8.8337
7	.12187	.99255	.12278	8.1443	1.0075	8.2055
7½	.13053	.99144	.13165	7.5957	1.0086	7.6613
8	.13917	.99027	.14054	7.1154	1.0098	7.1853
8½	.14781	.98901	.14945	6.6911	1.0111	6.7655
9	.15643	.98769	.15838	6.3137	1.0125	6.3924
9½	.16505	.98628	.16734	5.9758	1.0139	6.0588
10	.17365	.98481	.17633	5.6713	1.0154	5.7588
10½	.18223	.98325	.18534	5.3955	1.0170	5.4874
11	.19081	.98163	.19438	5.1445	1.0187	5.2408
11½	.19937	.97992	.20345	4.9151	1.0205	5.0158
12	.20791	.97815	.21256	4.7046	1.0223	4.8097
12½	.21644	.97630	.22169	4.5107	1.0243	4.6201
13	.22495	.97437	.23087	4.3315	1.0263	4.4454
13½	.23344	.97237	.24008	4.1653	1.0284	4.2836
14	.24192	.97029	.24933	4.0108	1.0306	4.1336
14½	.25038	.96815	.25862	3.8667	1.0329	3.9939
15	.25882	.96592	.26795	3.7320	1.0353	3.8637
15½	.26724	.96363	.27732	3.6059	1.0377	3.7420
16	.27564	.96126	.28674	3.4874	1.0403	3.6279
16½	.28401	.95882	.29621	3.3759	1.0429	3.5209
17	.29237	.95630	.30573	3.2708	1.0457	3.4203
17½	.30070	.95372	.31530	3.1716	1.0485	3.3255
18	.30902	.95106	.32492	3.0777	1.0515	3.2361

TRIGONOMETRIC FUNCTIONS *(cont.)*						
Degrees	Sin	Cos	Tan	Cot	Sec	Csc
18½	.31730	.94832	.33459	2.9887	1.0545	3.1515
19	.32557	.94552	.34433	2.9042	1.0576	3.0715
19½	.33381	.94264	.35412	2.8239	1.0608	2.9957
20	.34202	.93969	.36397	2.7475	1.0642	2.9238
20½	.35021	.93667	.37388	2.6746	1.0676	2.8554
21	.35837	.93358	.38386	2.6051	1.0711	2.7904
21½	.36650	.93042	.39391	2.5386	1.0748	2.7285
22	.37461	.92718	.40403	2.4751	1.0785	2.6695
22½	.38268	.92388	.41421	2.4142	1.0824	2.6131
23	.39073	.92050	.42447	2.3558	1.0864	2.5593
23½	.39875	.91706	.43481	2.2998	1.0904	2.5078
24	.40674	.91354	.44523	2.2460	1.0946	2.4586
24½	.41469	.90996	.45573	2.1943	1.0989	2.4114
25	.42262	.90631	.46631	2.1445	1.1034	2.3662
25½	.43051	.90258	.47697	2.0965	1.1079	2.3228
26	.43837	.89879	.48773	2.0503	1.1126	2.2812
26½	.44620	.89493	.49858	2.0057	1.1174	2.2411
27	.45399	.89101	.50952	1.9626	1.1223	2.2027
27½	.46175	.88701	.52057	1.9210	1.1274	2.1657
28	.46947	.88295	.53171	1.8807	1.1326	2.1300
28½	.47716	.87882	.54295	1.8418	1.1379	2.0957
29	.48481	.87462	.55431	1.8040	1.1433	2.0627
29½	.49242	.87035	.56577	1.7675	1.1489	2.0308
30	.50000	.86603	.57735	1.7320	1.1547	2.0000
30½	.50754	.86163	.58904	1.6977	1.1606	1.9703
31	.51504	.85717	.60086	1.6643	1.1666	1.9416
31½	.52250	.85264	.61280	1.6318	1.1728	1.9139
32	.52992	.84805	.62487	1.6003	1.1792	1.8871
32½	.53730	.84339	.63707	1.5697	1.1857	1.8611
33	.54464	.83867	.64941	1.5399	1.1924	1.8361
33½	.55194	.83388	.66188	1.5108	1.1992	1.8118
34	.55919	.82904	.67451	1.4826	1.2062	1.7883
34½	.56641	.82413	.68728	1.4550	1.2134	1.7655
35	.57358	.81915	.70021	1.4281	1.2208	1.7434
35½	.58070	.81411	.71329	1.4019	1.2283	1.7220
36	.58778	.80902	.72654	1.3764	1.2361	1.7013

TRIGONOMETRIC FUNCTIONS *(cont.)*

Degrees	Sin	Cos	Tan	Cot	Sec	Csc
36½	.59482	.80386	.73996	1.3514	1.2440	1.6812
37	.60181	.79863	.75355	1.3270	1.2521	1.6616
37½	.60876	.79335	.76733	1.3032	1.2605	1.6427
38	.61566	.78801	.78128	1.2799	1.2690	1.6243
38½	.62251	.78261	.79543	1.2572	1.2778	1.6064
39	.62932	.77715	.80978	1.2349	1.2867	1.5890
39½	.63608	.77162	.82434	1.2131	1.2960	1.5721
40	.64279	.76604	.83910	1.1917	1.3054	1.5557
40½	.64945	.76041	.85408	1.1708	1.3151	1.5398
41	.65606	.75471	.86929	1.1504	1.3250	1.5242
41½	.66262	.74895	.88472	1.1303	1.3352	1.5092
42	.66913	.74314	.90040	1.1106	1.3456	1.4945
42½	.67559	.73728	.91633	1.0913	1.3563	1.4802
43	.68200	.73135	.93251	1.0724	1.3673	1.4663
43½	.68835	.72537	.94896	1.0538	1.3786	1.4527
44	.69466	.71934	.96569	1.0355	1.3902	1.4395
44½	.70091	.71325	.98270	1.0176	1.4020	1.4267
45	.70711	.70711	1.00000	1.0000	1.4142	1.4142
45½	.71325	.70091	1.01760	.98270	1.4267	1.4020
46	.71934	.69466	1.03550	.96569	1.4395	1.3902
46½	.72357	.68835	1.05380	.94896	1.4527	1.3786
47	.73135	.68200	1.07240	.93251	1.4663	1.3673
47½	.73728	.67559	1.09130	.91633	1.4802	1.3563
48	.74314	.66913	1.11060	.90040	1.4945	1.3456
48½	.74895	.66262	1.13030	.88412	1.5092	1.3352
49	.75471	.65606	1.15040	.86929	1.5242	1.3250
49½	.76041	.64945	1.17080	.85408	1.5398	1.3151
50	.76604	.64279	1.19170	.83910	1.5557	1.3054
50½	.77162	.63608	1.21310	.82434	1.5721	1.2960
51	.77715	.62932	1.23490	.80978	1.5890	1.2867
51½	.78261	.62251	1.25720	.79543	1.6064	1.2778
52	.78801	.61566	1.27990	.78128	1.6243	1.2690
52½	.79335	.60876	1.30320	.76733	1.6427	1.2605
53	.79863	.60181	1.32700	.75355	1.6616	1.2521
53½	.80386	.59482	1.35140	.73996	1.6812	1.2440
54	.80902	.58778	1.37640	.72654	1.7013	1.2361

TRIGONOMETRIC FUNCTIONS (cont.)

Degrees	Sin	Cos	Tan	Cot	Sec	Csc
54½	.81411	.58070	1.4019	.71329	1.7220	1.2283
55	.81915	.57358	1.4281	.70021	1.7434	1.2208
55½	.82413	.56641	1.4550	.68728	1.7655	1.2134
56	.82904	.55919	1.4826	.67451	1.7883	1.2062
56½	.83388	.55194	1.5108	.66188	1.8118	1.1992
57	.83867	.54464	1.5399	.64941	1.8361	1.1924
57½	.84339	.53730	1.5697	.63707	1.8611	1.1857
58	.84805	.52992	1.6003	.62487	1.8871	1.1792
58½	.85264	.52250	1.6318	.61280	1.9139	1.1728
59	.85717	.51504	1.6643	.60086	1.9416	1.1666
59½	.86163	.50754	1.6977	.58904	1.9703	1.1606
60	.86603	.50000	1.7320	.57735	2.0000	1.1547
60½	.87035	.49242	1.7675	.56577	2.0308	1.1489
61	.87462	.48481	1.8040	.55431	2.0627	1.1433
61½	.87882	.47716	1.8418	.54295	2.0957	1.1379
62	.88295	.46947	1.8807	.53171	2.1300	1.1326
62½	.88701	.46175	1.9210	.52057	2.1657	1.1274
63	.89101	.45399	1.9626	.50952	2.2027	1.1223
63½	.89493	.44620	2.0057	.49858	2.2411	1.1174
64	.89879	.43837	2.0503	.48773	2.2812	1.1126
64½	.90258	.43051	2.0965	.47697	2.3228	1.1079
65	.90631	.42262	2.1445	.46631	2.3662	1.1034
65½	.90996	.41469	2.1943	.45573	2.4114	1.0989
66	.91354	.40674	2.2460	.44523	2.4586	1.0946
66½	.91706	.39875	2.2998	.43481	2.5078	1.0904
67	.92050	.39073	2.3558	.42447	2.5593	1.0864
67½	.92388	.38268	2.4142	.41421	2.6131	1.0824
68	.92718	.37461	2.4751	.40403	2.6695	1.0785
68½	.93042	.36650	2.5386	.39391	2.7285	1.0748
69	.93358	.35837	2.6051	.38386	2.7904	1.0711
69½	.93667	.35021	2.6746	.37388	2.8554	1.0676
70	.93969	.34202	2.7475	.36397	2.9238	1.0642
70½	.94264	.33381	2.8239	.35412	2.9957	1.0608
71	.94552	.32557	2.9042	.34433	3.0715	1.0576
71½	.94832	.31730	2.9887	.33459	3.1515	1.0545
72	.95106	.30902	3.0777	.32492	3.2361	1.0515

TRIGONOMETRIC FUNCTIONS (cont.)

Degrees	Sin	Cos	Tan	Cot	Sec	Csc
72½	.95372	.30070	3.1716	.31530	3.3255	1.0485
73	.95630	.29237	3.2708	.30573	3.4203	1.0457
73½	.95882	.28401	3.3759	.29621	3.5209	1.0429
74	.96126	.27564	3.4874	.28674	3.6279	1.0403
74½	.96363	.26724	3.6059	.27732	3.7420	1.0377
75	.96592	.25882	3.7320	.26795	3.8637	1.0353
75½	.96815	.25038	3.8667	.25862	3.9939	1.0329
76	.97029	.24192	4.0108	.24933	4.1336	1.0306
76½	.97237	.23344	4.1653	.24008	4.2836	1.0284
77	.97437	.22495	4.3315	.23087	4.4454	1.0263
77½	.97630	.21644	4.5107	.22169	4.6201	1.0243
78	.97815	.20791	4.7046	.21256	4.8097	1.0223
78½	.97992	.19937	4.9151	.20345	5.0158	1.0205
79	.98163	.19081	5.1445	.19438	5.2408	1.0187
79½	.98325	.18223	5.3955	.18534	5.4874	1.0170
80	.98481	.17365	5.6713	.17633	5.7588	1.0154
80½	.98628	.16505	5.9758	.16734	6.0588	1.0139
81	.98769	.15643	6.3137	.15838	6.3924	1.0125
81½	.98901	.14781	6.6911	.14945	6.7655	1.0111
82	.99027	.13917	7.1154	.14054	7.1853	1.0098
82½	.99144	.13053	7.5957	.13165	7.6613	1.0086
83	.99255	.12187	8.1443	.12278	8.2055	1.0075
83½	.99357	.11320	8.7769	.11393	8.8337	1.0065
84	.99452	.10453	9.5144	.10510	9.5668	1.0055
84½	.99540	.09584	10.3850	.09629	10.4330	1.0046
85	.99619	.08715	11.4300	.08749	11.4740	1.0038
85½	.99692	.07846	12.7060	.07870	12.7450	1.0031
86	.99756	.06976	14.3010	.06993	14.3350	1.0024
86½	.99813	.06105	16.3500	.06116	16.3800	1.0019
87	.99863	.05234	19.0810	.05241	19.1070	1.0014
87½	.99905	.04362	22.9040	.04366	22.9250	1.0009
88	.99939	.03490	28.6360	.03492	28.6540	1.0006
88½	.99966	.02618	38.1880	.02618	38.2010	1.0003
89	.99985	.01745	57.2900	.01745	57.2990	1.0001
89½	.99996	.00873	114.5900	.00873	114.5900	1.0000
90	1.00000	0	Inf.	0	Inf.	1.0000

CHAPTER 3
Design Data

WATER BASICS
1 ft.^{3} of water contains 7.48 gal., 1,728 in.^{3}, and weighs 62.48 lb.
1 ft.^{3} of ice weighs 57.2 lb.
1 gal. of water weighs 8.33 lb. and contains 231 in.^{3} or 0.1337 ft.^{3}
1 lb. of water equals 27.72 in.^{3}
One BTU is the heat needed to raise one pound of water one degree F.
1 ft. of water equals 0.434 psi
2.31 ft. of water equals 1.0 psi
The height of a column of water, equal to a pressure of 1.0 psi, is 2.31 ft.
To find the pressure in psi of a column of water, multiply the height of the column in feet by 0.434.
The average pressure of the atmosphere is estimated at 14.7 psi so that with a perfect vacuum it will sustain a column of water 34 ft. high.
Water expands $\frac{1}{23}$ of its volume when heated from 40° to 212°F.
Water is at its greatest density at 39.2°F
The friction of water in pipes varies as the square of the velocity.
To evaporate 1 ft.^{3} of water requires the consumption of $7\frac{1}{2}$ lb. of ordinary coal or about 1 lb. of coal to 1 gal. of water.
1 in.^{3} of water evaporated at atmospheric pressure is converted into approximately 1 ft.^{3} of steam.

METRIC LIQUID VOLUME EQUIVALENTS

Metric	U.S.
3.7854 L	1 gallon
0.946 L	1 quart
0.473 L	1 pint
1 L	0.264 gallons
1 L	33.814 ounces
29.576 ml	1 fluid ounce
236.584 ml	1 cup

METRIC LENGTH EQUIVALENTS

Metric	U.S.
1 rn	39.37 inches
1 m	3.28 feet
1 m	1.094 yards
1 m	.0016 mile
1 km	0.625 miles
1.609 km	1 mile
25.4 mm	1 inch
2.54 cm	1 inch
304.8 mm	1 foot
1 mm	0.03937 inch
1 cm	0.3937 inch
1 dm	3.937 inches

METRIC PRESSURE CONVERSIONS

Measurement	Equivalent
1 pound per square inch (psi)	6.8947 kPa
1 m column of water	9.794 kPa
10.2 cm of water	1 kPa
1 cm column of mercury	1.3332 kPa
1 inch of mercury (inHg)	3.3864 kPa
6 cm of mercury	8 kPa

PRESSURE

The kilopascal (kPa) is the unit of measurement recommended for fluid pressure.

When working with pressure, note that atmospheric pressure is measured at 101.3 kPa metric and 14.7 psi English. Plumbers need to calculate pressures for columns of water. The following facts will help you to understand pressures measured in kilopascals.

- Atmospheric pressure of 101.3 kPa will support a column of mercury 76 cm high.
- To find head pressure in decimeters when pressure is given in kPa, divide pressure by 0.9794.
- To find head pressure in kPa of a column of water given in decimeters, multiply the number of decimeters by 0.9794.
- 1 kilopascal (kPa) of air pressure elevates water approximately 10.2 cm under atmospheric conditions of 101 kPa.
- 10.2 cm of water equals 1 kPa.
- 51 cm of water equals 5 kPa.
- 1 meter of water equals 9.8 kPa.
- 10,000 square centimeters equals 1 square meter.
- 1 cubic meter equals 1,000,000 cubic centimeters (cm^3) or 1000 cubic decimeters (dm^3).
- 1 liter of cold water weighs 1 kilogram.

AIR PRESSURE

- 1 cubic inch of mercury weighs 0.49 lbs.
- Generally speaking, 2 inches of mercury is equivalent to 1 pound of pressure (psi).
- 1 cubic foot of air weighs 0.075 lbs.
- 1 cubic meter of air weighs 1.214 kilograms.

ABSOLUTE ZERO

- Absolute zero is -459.69°F.
- Absolute zero is -273.16°C.

WATER FORMULAS

Gallons per minute through a pipe:

GPM = 0.0408 x [Pipe I.D. (in.)]^2 x Feet/Water Velocity(min.)

Horsepower to raise water:

$$HP = \frac{Total\ Head\ (ft.)\ x\ GPM}{3960}$$

Note: For non-water liquids, multiply GPM by the liquid's specific gravity.

Round Tank capacity in gallons:

Round Tank (gal.) =

[Tank Dia. (ft.)]^2 x 0.7854 x Tank Length (ft.) x 7.48 (gal. per ft.^{3})

Weight of water in a pipe:

Water (lb.) = 0.34 x [Pipe I.D. (in.)]^2 x Pipe Length (ft.)

BOILING POINTS OF WATER AT SEA LEVEL

Gauge (psi)	Boiling Point (F)
0	212°
2	218.5°
4	224.4°
6	229.8°
8	234.8°
10	239.4°
15	249.8°
25	266.8°
50	297.1°
100	337.9°
125	352.9°
200	387.9°

Note: A 0 gauge pressure equals 14.7 psi actual air pressure at sea level.

WEIGHT OF WATER IN POUNDS					
Temp. (F)	Wt. Per Cu. Ft.	Wt. Per Gallon	Temp. (F)	Wt. Per Cu. Ft.	Wt. Per Gallon
32°	62.418	8.344	130°	61.563	8.230
35°	62.422	8.345	135°	61.472	8.218
39.2°	62.425	8.346	140°	61.381	8.206
40°	62.425	8.346	145°	61.291	8.193
45°	62.422	8.345	150°	61.201	8.181
50°	62.409	8.343	155°	61.096	8.167
55°	62.394	8.341	160°	60.991	8.153
60°	62.372	8.338	165°	60.843	8.134
65°	62.344	8.334	170°	60.783	8.126
70°	62.313	8.331	175°	60.665	8.110
75°	62.275	8.325	180°	60.548	8.094
80°	62.232	8.321	185°	60.430	8.078
85°	62.182	8.313	190°	60.314	8.063
90°	62.133	8.306	195°	60.198	8.047
95°	62.074	8.297	200°	60.081	8.032
100°	62.022	8.291	205°	59.980	8.018
105°	61.960	8.283	210°	59.820	7.997
110°	61.868	8.271	212°	59.760	7.989
115°	61.807	8.261	250°	58.750	7.854
120°	61.715	8.250	300°	56.970	7.616
125°	61.654	8.242	400°	54.250	7.252

PIPE CONTENTS			
	Per 1 Foot of Length		
Nominal Pipe Size (I.D.) in Inches	Cubic Inches	Gallons	Weight of Water in Pounds
¼	.5891	.0026	.0213
⅜	1.325	.0057	.0478
½	2.356	.0102	.0850
¾	5.301	.0229	.1913
1	9.425	.0408	.3400
1¼	14.726	.0637	.5313
1½	21.205	.0918	.7650
2	37.699	.1632	1.3600
2½	58.905	.2550	2.1250
3	84.823	.3672	3.0600
4	150.797	.6528	5.4400
5	235.620	1.020	8.5000
6	339.293	1.469	12.2400
8	603.187	2.611	21.7600
10	942.480	4.080	34.0000
12	1357.171	5.875	48.9600
15	2120.580	9.180	76.5000

FLOW-RATE CONVERSIONS

Gallons per Minute	Liters per Minute
1	3.75
2	6.50
3	11.25
4	15.00
5	18.75
6	22.50
7	26.25
8	30.00
9	33.75
10	37.50

Feet per Second	Meters per Second
1	.305
2	.610
3	.915
4	1.220
5	1.525
6	1.830
7	2.135
8	2.440
9	2.754
10	3.050

FLOW-RATE EQUIVALENTS

1 gal. per min. (gpm) = 0.1337 cu. ft. per min. (cfm)

1 cu. ft. per min. (cfm) = 448.8 gal. per hr. (gph)

WATER DEMAND AT INDIVIDUAL OUTLETS

Outlet	gpm
Ordinary lavatory faucet	2.00
Bath faucet, ½ in.	5.00
Laundry faucet, ½ in.	5.00
Self-closing lavatory faucet	2.50
Sink faucet, ⅜ or ½ in.	4.50
Sink faucet, ¾ in.	6.00
Shower head, ½ in.	5.00
Ballcock in water closet flush tank	3.00
¾ in. flush valve (15 psi)	15.00
1 in. flush valve (15 psi)	27.00
1 in. flush valve (25 psi)	35.00
Drinking fountain jet	.75
Dishwashing machine (domestic)	4.00
Laundry machine, 8 or 16 lb.	4.00
Aspirator (laboratory or operating room)	2.50
Hose bib or sillcock, ½ in.	5.00

AVERAGE WATER FLOW PER FIXTURE	
Fixture	**Flow Rate (gpm)**
Ordinary basin faucet	2.00
Self-closing basin faucet	2.50
Sink faucet, ⅜ in.	4.50
Sink faucet, ½ in.	4.50
Bathtub faucet	6.00
Shower	5.00
Laundry tub cock, ½ in.	5.00
Ballcock for water closet	3.00
Flushometer valve for water closet	15.00 to 35.00
Flushometer valve for urinal	15.00
Drinking fountain	.75
Sillcock (wall hydrant)	5.00
AVERAGE WATER USAGE PER ACTIVITY	
Activity	**Gallons Used**
Brushing teeth	1.00
Washing hands	2.00
Shower	25.00
Tub bath	36.00
Dishwashing	50.00
Automatic dishwasher cycle	16.00
Washing machine cycle	60.00

FIXTURE-UNIT RATINGS	
Fixture	**Rating**
Lavatory	1
Bathtub	2
Shower	2
Residential toilet	4
Kitchen sink	2
Dishwasher	2
Laundry tub	2
Washing machine	3

EQUIVALENT FIXTURE-UNIT RATINGS	
Fixture	**Hot and Cold Water Combined**
Lavatory	1
Tub/shower combination	2
Shower	2
Toilet	3
Dishwasher	2
Kitchen sink	2
Laundry hookup	2
Sillcock	3

PRESSURE AND FLOW RATE PER FIXTURE

Fixture	Pressure (psi)	Flow (gpm)
Ordinary basin faucet	8	2.0
Self-closing basin faucet	12	2.5
Sink faucet, 3/8"	10	4.5
Sink faucet, 1/2"	8	5.0
Bathtub faucet	8	4.0
Shower	8	2.0
Laundry tub cock, 1/2"	8	4.0
Ballcock for water closet	8	3.0
Flush valve for water closet	15	25.0
Flush valve for urinal	15	15.0
50' garden hose and sillcock	8	5.0

BRANCH PIPING FOR FIXTURES

Fixture	Minimum NPS (in.)
Bathtub	$\frac{1}{2}$
Combination sink and laundry tray	$\frac{1}{2}$
Drinking fountain	$\frac{3}{8}$
Dishwashing machine (domestic)	$\frac{1}{2}$
Kitchen sink (domestic)	$\frac{1}{2}$
Kitchen sink (commercial)	$\frac{3}{4}$
Lavatory	$\frac{3}{8}$
Laundry tray (1 to 3 compartments)	$\frac{1}{2}$
Shower (single head)	$\frac{1}{2}$
Sink (service, slop)	$\frac{1}{2}$
Sink (flushing rim)	$\frac{3}{4}$
Urinal (flush tank)	$\frac{1}{2}$
Urinal ($\frac{3}{4}$ in. flush valve)	$\frac{3}{4}$
Urinal (1 in. flush valve)	1
Water closet (flush tank)	$\frac{3}{8}$
Water closet (flush valve)	1
Hose bib	$\frac{1}{2}$
Wall hydrant or sillcock	$\frac{1}{2}$

SERVICE AND DISTRIBUTION PIPE SIZING (46 TO 60 PSI)

Size of Water Meter and Street Service NPS (in.)	Size of Water Service and Distribution Pipes NPS (in.)	Maximum Length of Water Pipe in feet					
		Number of Fixture Units					
		40'	60'	80'	100'	150'	200'
¾	½	9	8	7	6	5	4
¾	¾	27	23	19	17	14	11
¾	1	44	40	36	33	28	23
1	1	60	47	41	36	30	25
1	1¼	102	87	76	67	52	44

MINIMUM SIZE FOR FIXTURE SUPPLY

Fixture	Minimum NPS (in.)
Lavatory	$\frac{3}{8}$
Bidet	$\frac{3}{8}$
Toilet	$\frac{3}{8}$
Bathtub	$\frac{1}{2}$
Shower	$\frac{1}{2}$
Kitchen sink	$\frac{1}{2}$
Dishwasher	$\frac{1}{2}$
Laundry tub	$\frac{1}{2}$
Hose bib	$\frac{1}{2}$

INDIVIDUAL, BRANCH AND CIRCUIT VENT SIZING FOR HORIZONTAL DRAIN PIPES

Drain Pipe Size (in.)	Drain Pipe Grade per Foot (in.)	Vent Pipe Size (in.)	Maximum Developed Length of Vent Pipe (ft.)
$1\frac{1}{2}$	$\frac{1}{4}$	$1\frac{1}{4}$	Unlimited
$1\frac{1}{2}$	$\frac{1}{4}$	$1\frac{1}{2}$	Unlimited
2	$\frac{1}{4}$	$1\frac{1}{4}$	290
2	$\frac{1}{4}$	$1\frac{1}{2}$	Unlimited
3	$\frac{1}{4}$	$1\frac{1}{2}$	97
3	$\frac{1}{4}$	2	420
3	$\frac{1}{4}$	3	Unlimited
4	$\frac{1}{4}$	2	98
4	$\frac{1}{4}$	3	Unlimited
4	$\frac{1}{4}$	4	Unlimited

DETERMINING SIZE AND LENGTHS OF VENT STACKS

Size of Soil or Waste Stack (In.)	Fixture Units Connected (dfu)	Vent Diameter (in.)										
		1¼	1½	2	2½	3	4	5	6	8	10	12
		Maximum Developed Length of Vent in Feet										
1¼	2	50	–	–	–	–	–	–	–	–	–	–
1½	4	40	200	–	–	–	–	–	–	–	–	–
2	9	–	100	200	–	–	–	–	–	–	–	–
2	18	–	50	150	–	–	–	–	–	–	–	–
2½	42	–	30	100	300	–	–	–	–	–	–	–
3	72	–	–	50	80	400	–	–	–	–	–	–
4	240	–	–	40	70	250	–	–	–	–	–	–
4	500	–	–	–	50	180	700	–	–	–	–	–
5	540	–	–	–	–	150	600	–	–	–	–	–
5	1100	–	–	–	–	50	200	700	–	–	–	–
6	1900	–	–	–	–	–	50	200	700	–	–	–
8	2200	–	–	–	–	–	–	150	500	–	–	–
8	3600	–	–	–	–	–	–	60	250	800	–	–
10	3800	–	–	–	–	–	–	–	200	600	–	–
10	5600	–	–	–	–	–	–	–	60	250	800	–
12	6000	–	–	–	–	–	–	–	–	200	600	–
12	8400	–	–	–	–	–	–	–	–	100	300	900
15	10500	–	–	–	–	–	–	–	–	50	200	600
15	50000	–	–	–	–	–	–	–	–	–	75	180

VENT STACK AND STACK VENT SIZING

Drain Pipe Size (in.)	Fixture-Unit Load on Drain Pipe	Vent Pipe Size (in.)	Maximum Developed Length of Vent Pipe (ft.)
1½	8	1¼	50
1½	8	1½	150
1½	10	1¼	30
1½	10	1½	100
2	12	1½	75
2	12	2	200
2	20	1½	50
2	20	2	150
3	10	1½	42
3	10	2	150
3	10	3	1040
3	21	1½	32
3	21	2	110
3	21	3	810
3	102	1½	25
3	102	2	86
3	102	3	620
4	43	2	35
4	43	3	250
4	43	4	980
4	540	2	21
4	540	3	150
4	540	4	580

WET-VENT STACK SIZING	
Quantity of Wet-Vented Fixtures	**Required Stack Size (in.)**
1 to 2 Bathtubs or showers	2
3 to 5 Bathtubs or showers	2½
6 to 9 Bathtubs or showers	3
10 to 16 Bathtubs or showers	4

WET STACK VENT SIZING		
Stack Size (in.)	**Fixture-Unit Load on Stack**	**Maximum Stack Length (ft.)**
2	4	30
3	24	50
4	50	100
6	100	300

VENT SIZE	
Fixture	**Minimum Vent Size (in.)**
Bathtub	1¼
Lavatory	1¼
Domestic sink	1¼
Shower stalls, domestic	1¼
Laundry tray	1¼
Drinking fountain	1¼
Service sink	1¼
Water closet	2

A minimum of one 3 in. vent must be installed.

TRAP-TO-VENT DISTANCES			
Grade on Drain Pipe (in.)	**Drain Size (in.)**	**Trap Size (in.)**	**Maximum Distance Between Trap and Vent (ft.)**
¼	1¼	1¼	3½
¼	1½	1¼	5
¼	1½	1½	5
¼	2	1½	8
¼	2	2	9
⅛	3	3	10
⅛	4	4	12

TRAP SIZES

Appliance or Fixture	Size (in.)
Lavatory	$1\frac{1}{4}$
Drinking fountain	$1\frac{1}{4}$
Dental unit or cuspidor	$1\frac{1}{4}$
Bathtub with or without shower	$1\frac{1}{2}$
Bidet	$1\frac{1}{2}$
Laundry tray	$1\frac{1}{2}$
Dishwasher, domestic	$1\frac{1}{2}$
Dishwasher, commercial	2
Washing machine	2
Floor drain	2, 3 or 4
Shower stall, domestic	2
Sinks:	
Combination, sink and tray (with disposal unit)	$1\frac{1}{2}$
Combination, sink and tray (with one trap)	$1\frac{1}{2}$
Domestic, with or without disposal unit	$1\frac{1}{2}$
Commercial, flat rim, bar or counter	$1\frac{1}{2}$
Circular or multiple wash	$1\frac{1}{2}$
Soda fountain	$1\frac{1}{2}$
Surgical	$1\frac{1}{2}$
Laboratory	$1\frac{1}{2}$
Pot or scullery	2
Service sink	2 or 3
Flushrim or bedpan washer	3
Urinals:	
Trough (per 6 ft. section)	$1\frac{1}{2}$
Wall-hung	$1\frac{1}{2}$ or 2
Stall	2
Pedestal	3
Water closet	3

SIZING COPPER AND GALVANIZED STEEL SERVICE PIPE

Number of bathrooms and kitchens with Tank-type closets using		Size of water service pipe (in.)	Recommended meter size (in.)	Approximate pressure loss for meter and 100' of pipe (psi)	Number of bathrooms and kitchens with Flush valve closets using	
Copper	Galvanized				Copper	Galvanized
1-2	–	3/4	5/8	27	–	–
–	1-2	3/4	5/8	40	–	–
–	–	1	1	30	1	–
3-4	–	1	1	22	–	–
–	3-4	1	1	24	–	–
–	–	1¼	1	32	2-3	–
–	–	1¼	1	36	–	1-2
5-9	–	1¼	1	28	–	–
–	5-8	1¼	1	32	–	–
–	–	1½	1½	29	4-10	–
–	–	1½	1½	30	–	3-7
–	–	1½	1½	17	–	–
–	9-14	1½	1½	21	–	–
10-16	–	2	1½	26	11-18	–
–	–	2	1½	32	–	8-18

SIZING COPPER AND GALVANIZED STEEL SERVICE PIPE (cont.)						
Number of bathrooms and kitchens with Tank-type closets using		Size of water service pipe (in.)	Recommended meter size (in.)	Approximate pressure loss for meter and 100' of pipe (psi)	Number of bathrooms and kitchens with Flush valve closets using	
Copper	Galvanized				Copper	Galvanized
17-38	–	2	1½	27	–	–
–	15-38	2	1½	32	–	–
–	–	2	2	25	19-33	–
–	–	2	2	24	–	19-24
39-56	–	2	2	25	–	–
–	39-45	2	2	24	–	–
–	–	2½	2	28	34-57	–
–	–	2½	2	32	–	25-57
57-78	–	2½	2	28	–	–
–	46-78	2½	2	32	–	–
–	–	3	3	16	58-95	–
–	–	3	3	19	–	58-95
79-120	–	3	3	16	–	–
–	79-120	3	3	19	–	–

HOT WATER DEMAND PER FIXTURE IN GALLONS PER HOUR AT 140°F

	Apt. house	Private home	Club	Gym	Office bldg.	Hospital	Hotel	Indus. plant	School	Y.M.C.A.
Basins, private lavatory	2	2	2	2	2	2	2	2	2	2
Basins, public lavatory	4	–	6	8	6	6	8	12	15	8
Bathtubs	20	20	20	30	–	20	20	–	–	30
Dishwashers	15	15	50-150	–	–	50-150	50-200	20-100	20-100	20-100
Foot basins	3	3	3	12	–	3	3	12	3	12
Kitchen sink	10	10	20	–	20	20	30	20	20	20
Laundry, stationary tubs	20	20	28	–	–	28	28	–	–	28
Pantry sink	5	5	10	–	10	10	10	–	10	10
Showers	30	30	150	225	30	75	75	225	225	225
Service sink	20	15	20	–	20	20	30	20	20	20
Hydro-therapeutic showers	–	–	–	–	–	400	–	–	–	–
Hubbard baths	–	–	–	–	–	600	–	–	–	–
Leg baths	–	–	–	–	–	100	–	–	–	–
Arm baths	–	–	–	–	–	35	–	–	–	–
Sitz baths	–	–	–	–	–	30	–	–	–	–
Continuous-flow baths	–	–	–	–	–	165	–	–	–	–
Circular wash sinks	–	–	–	–	20	20	20	30	30	–
Semi-circular wash sinks	–	–	–	–	10	10	10	15	15	–

3-22

RECOMMENDED MAXIMUM REQUIREMENTS FOR HOT WATER PER DAY IN GALLONS AT 140°F

Type of structure	Number of rooms	Number of bathrooms				
		1	2	3	4	5
	1	60	–	–	–	–
	2	70	–	–	–	–
	3	80	–	–	–	–
	4	90	120	–	–	–
	5	100	140	–	–	–
	6	120	160	200	–	–
Apartments,	7	140	180	220	–	–
condominiums,	8	160	200	240	250	–
townhouses	9	180	220	260	275	–
and	10	200	240	280	300	–
private homes	11	–	260	300	340	–
	12	–	280	325	380	450
	13	–	300	350	420	500
	14	–	–	375	460	550
	15	–	–	400	500	600
	16	–	–	–	540	650
	17	–	–	–	580	700
	18	–	–	–	620	750
	19	–	–	–	–	800
	20	–	–	–	–	850
Hotels	Room with basin	10				
	Room with bath — transient	50				
	Room with bath — resident	60				
	2 rooms with bath	80				
	3 rooms with bath	100				
Public facilities	Public shower	200				
	Public basins	150				
	Slop sink	30				
Office buildings	Office staff (per person)	2.3				
	Other personnel (per person)	4.0				
	Cleaning per 10,000 sq. ft.	30.0				
Hospitals	Per bed	80-100				

RECOMMENDED HOT WATER TEMPERATURES

Usage	Temperature (F)
Lavatory, hand washing	105°
Lavatory, face washing	115°
Shower and bathtub	110°
Dishwasher and laundry, private	140°
Dishwasher, washing	140°
Dishwasher, sanitizing	180°
Laundry, commercial and institutional	180°
Surgical scrubbing	110°
Commercial buildings, occupant use	110°

MINIMUM RECOMMENDED HOT WATER STORAGE TANK CAPACITY—RESIDENTIAL

Heater Fuel type	Total number of bedrooms	Storage tank capacity (gal.)
Gas		20
Electric	1	30
Oil		30
Gas		30
Electric	2	40
Oil		30
Gas		30
Electric	3	50
Oil		30
Gas		40
Electric	4	66
Oil		30

STORAGE TANK DIMENSIONS WITH CAPACITIES IN GALLONS

Tank Length in Ft.	Tank Diameter in Inches										
	20"	22"	24"	30"	36"	42"	48"	54"	60"
1	16	20	24	37	53	72	94	120	145
2	32	40	48	74	106	144	188	240	290
3	48	60	72	110	159	216	282	360	435
4	66	80	96	147	212	288	376	480	580
5	82	100	120	184	265	360	470	600	725
6	98	120	144	220	317	432	564	720	870
7	114	140	168	257	370	504	658	840	1015
8	131	160	192	294	423	576	752	960	1160
9	147	180	216	330	476	648	846	1080	1305
10	163	200	240	367	529	720	940	1200	1450

NATURAL GAS PIPE CAPACITIES IN CFM

Pipe Size (in.)	Pipe Length in Feet						
	10	20	30	40	50	60	70
½	170	118	95	80	71	64	60
¾	360	245	198	169	150	135	123
1	670	430	370	318	282	255	235
1¼	1320	930	740	640	565	510	470
1½	1990	1370	1100	950	830	760	700
2	3880	2680	2150	1840	1610	1480	1350
2½	6200	4120	3420	2950	2600	2360	2180
3	10900	7500	6000	5150	4600	4150	3820
3½	16000	11000	8900	7600	6750	6200	5650
4	22500	15500	12400	10600	9300	8500	7900

Pipe Size (in.)	Pipe Length in Feet						
	80	90	100	125	150	200	250
½	55	52	49	44	40	34	30
¾	115	108	102	92	83	71	63
1	220	205	192	172	158	132	118
1¼	440	410	390	345	315	270	238
1½	650	610	570	510	460	400	350
2	1250	1180	1100	1000	910	780	690
2½	2000	1900	1800	1600	1450	1230	1100
3	3550	3300	3120	2810	2550	2180	1930
3½	5250	4950	4650	4150	3800	3200	2860
4	7300	6800	6400	5700	5200	4400	3950

Pipe Size (in.)	Pipe Length in Feet						
	300	350	400	450	500	550	600
½	27	25	23	22	21	20	19
¾	57	52	48	45	43	41	39
1	108	100	92	86	81	77	74
1¼	215	200	185	172	162	155	150
1½	320	295	275	255	240	230	220
2	625	570	535	500	470	450	430
2½	1000	920	850	800	760	720	690
3	1750	1600	1500	1400	1320	1250	1200
3½	2600	2400	2200	2100	2000	1900	1800
4	3600	3250	3050	2850	2700	2570	2450

APPROXIMATE GAS DEMAND FOR COMMON APPLIANCES

Appliance	Input-Btuh
Commercial kitchen:	
Small broiler	30,000
Large broiler	60,000
Combination broiler and roaster	66,000
2 deck baking and roasting oven	100,000
3 deck baking oven	96,000
Deep fat fryer, 45 pound capacity	50,000
Deep fat fryer, 75 pound capacity	75,000
Doughnut fryer, 200 pound capacity	72,000
Revolving oven, 4 or 5 trays	210,000
Range with hot top	45,000
Range with hot top and oven	90,000
Range with fry top	50,000
Range with fry top and oven	100,000
Coffee maker, 3 burner	18,000
Coffee maker, 4 burner	24,000
Coffee urn, single, 5 gal. capacity	28,000
Coffee urn, twin, 10 gal. capacity	56,000
Coffee urn, twin, 15 gal. capacity	84,000
Residential:	
Oven	25,000
Stove top burners	40,000
Range	65,000
Clothes dryer	35,000
Water heater, 30 gal. capacity	30,000
Water heater, 40 to 50 gal. capacity	50,000
Log lighter, residential	25,000
Barbeque/grill	50,000
Miscellaneous:	
Gas engine, per horsepower	10,000
Bunsen burner	30,000
Log lighter, commercial	50,000
Steam boiler, per horsepower	50,000

TROUBLESHOOTING – WATER HEATERS

Problem	Analysis	Action
Condensation	Heater is installed in a closed or confined area	Vent room or install louvers to allow more air circulation
High operating costs	Lime, rust or sediment in tank	Drain and flush (replace heater if buildup is severe)
	Heater too small for application	Replace with larger heater
	Wrong size piping connections	Install correct size piping
	Lack of insulation on piping	Insulate pipes
Relief valve drips	Excessive temperature setting	Lower setting on thermostat
	Defective valve	Replace valve
Rusty hot water	Buildup of rust in heater	Drain and flush (replace heater if buildup is severe)
Water tank leaks	Rusting of inner tank walls causing pin holes	Replace heater (repairs are only temporary)

TROUBLESHOOTING – ELECTRIC WATER HEATERS

Problem	Analysis	Action
Extremely hot water comes out of faucet	Defective thermostat	Replace thermostat
	Excessive temperature setting	Lower setting on thermostat
Hot water turns cold quickly	Defective upper or lower heating elements	Replace any defective elements
	Defective thermostat	Replace thermostat
	Rust buildup	Drain and flush (replace heater if buildup is severe)
No hot water	Blown fuses or defective circuit breaker	Replace fuses adjust or replace circuit breaker
	Defective heating element(s)	Replace element(s)
	Defective thermostat	Replace thermostat
	Defective time clock	Replace time clock

TROUBLESHOOTING – GAS WATER HEATERS

Problem	Analysis	Action
Extremely hot water	Defective gas control valve	Replace gas control valve
	Defective thermostat	Replace thermostat
	Excessive temperature setting	Lower setting on thermostat
Gas smell is detected	Hole in fittings, connections, valves or flue	Replace any defective or worn parts; tighten loose connections
No gas	No gas coming in to burner	Check valves and adjust or replace
	Defective thermo-coupling	Replace thermo-coupling
	Dip tube installed incorrectly or defective	Reinstall or replace tube
Water turns cold quickly	Dip tube in wrong inlet or broken	Insert tube in cold water supply or replace tube
	Gas controller defective	Replace all defective parts

PRESSURE LOSS DUE TO FRICTION IN COPPER TUBE

Pressure Loss per 100 Feet of Tube (psi)

Flow (gpm)	Standard Type M Tube Size in Inches											
	3/8	1/2	3/4	1	1¼	1½	2	2½	3	4	5	6
1	2.5	0.8	0.2	–	–	–	–	–	–	–	–	–
2	8.5	2.8	0.5	0.2	–	–	–	–	–	–	–	–
3	17.3	5.7	1.0	0.3	0.1	–	–	–	–	–	–	–
4	28.6	9.4	1.8	0.5	0.2	–	–	–	–	–	–	–
5	42.2	13.8	2.6	0.7	0.3	0.1	–	–	–	–	–	–
10	–	46.6	8.6	2.5	0.9	0.4	0.1	–	–	–	–	–
15	–	–	17.6	5.0	1.9	0.9	0.2	–	–	–	–	–
20	–	–	29.1	8.4	3.2	1.4	0.4	0.1	–	–	–	–
25	–	–	–	12.3	4.7	2.1	0.6	0.2	–	–	–	–
30	–	–	–	17.0	6.5	2.9	0.8	0.3	0.1	–	–	–
35	–	–	–	–	8.5	3.8	1.0	0.4	0.2	–	–	–
40	–	–	–	–	11.0	4.9	1.3	0.5	0.2	–	–	–
45	–	–	–	–	13.6	6.1	1.6	0.6	0.2	–	–	–
50	–	–	–	–	–	7.3	2.0	0.7	0.3	–	–	–
60	–	–	–	–	–	10.2	2.7	1.0	0.4	–	–	–
70	–	–	–	–	–	13.5	3.6	1.2	0.5	0.1	–	–
80	–	–	–	–	–	–	4.6	1.6	0.7	0.2	–	–
90	–	–	–	–	–	–	5.7	2.0	0.9	0.2	–	–
100	–	–	–	–	–	–	7.5	2.7	1.0	0.3	0.1	–
200	–	–	–	–	–	–	–	8.5	3.6	1.0	0.3	0.1
300	–	–	–	–	–	–	–	–	8.0	2.0	0.7	0.3
400	–	–	–	–	–	–	–	–	–	3.3	1.2	0.5
500	–	–	–	–	–	–	–	–	–	–	1.7	0.7
750	–	–	–	–	–	–	–	–	–	–	3.6	1.5
1000	–	–	–	–	–	–	–	–	–	–	–	2.5

WATER FRICTION TABLES

⅜" Copper Tubing (S.P.S.)*

Water Flow (gpm)	Type K		Type L		Type M	
	.402" I.D. .049" wall		.430" I.D. .035" wall		.450" I.D. .025" wall	
	Water Velocity ft./sec.	Head loss in ft./100 ft.	Water Velocity ft./sec.	Head loss in ft./100 ft.	Water Velocity ft./sec.	Head loss in ft./100 ft.
0.2	.51	.66	.44	.48	.40	.39
0.4	1.01	2.15	.88	1.57	.81	1.27
0.6	1.52	4.29	1.33	3.12	1.21	2.52
0.8	2.02	7.02	1.77	5.11	1.61	4.12
1.0	2.52	10.32	2.20	7.50	2.01	6.05
1.5	3.78	20.86	3.30	15.15	3.02	12.21
2.0	5.04	34.48	4.40	20.03	4.02	20.16
2.5	6.30	51.03	5.50	37.01	5.03	29.80
3.0	7.55	70.38	6.60	51.02	6.04	41.07
3.5	8.82	92.44	7.70	66.98	7.04	53.90
4.0	10.10	117.10	8.80	84.85	8.05	68.26
4.5	11.40	144.40	9.90	104.60	9.05	84.11
5.0	12.60	174.30	11.00	126.10	10.05	101.40

*A safety factor of 15-20% should be added to these values.

WATER FRICTION TABLES (cont.)

½" Copper Tubing (S.P.S.)*

Water Flow (gpm)	Type K		Type L		Type M	
	.527" I.D. .049" wall		.545" I.D. .040" wall		.569" I.D. .028" wall	
	Water Velocity ft./sec.	Head loss in ft./100 ft.	Water Velocity ft./sec.	Head loss in ft./100 ft.	Water Velocity ft./sec.	Head loss in ft./100 ft.
0.5	.74	.88	.69	.75	.63	.62
1.0	1.47	2.87	1.38	2.45	1.26	2.00
1.5	2.20	5.77	2.06	4.93	1.90	4.02
2.0	2.94	9.52	2.75	8.11	2.53	6.61
2.5	3.67	14.05	3.44	11.98	3.16	9.76
3.0	4.40	19.34	4.12	16.48	3.79	13.42
3.5	5.14	25.36	4.81	21.61	4.42	17.59
4.0	5.87	32.09	5.50	27.33	5.05	22.25
4.5	6.61	39.51	6.19	33.65	5.68	27.39
5.0	7.35	47.61	6.87	40.52	6.31	32.99
6.0	8.81	65.79	8.25	56.02	7.59	45.57
7.0	10.30	86.57	9.62	73.69	8.84	59.93
8.0	11.80	109.90	11.00	93.50	10.10	76.03
9.0	13.20	135.60	12.40	115.40	11.40	93.82
10.0	14.70	163.80	13.80	139.40	12.60	113.30

*A safety factor of 15-20% should be added to these values.

WATER FRICTION TABLES (cont.)

5⁄8" Copper Tubing (S.P.S.)*

Water Flow (gpm)	Type K		Type L		Type M	
	.652" I.D. .049" wall		.666" I.D. .042" wall		.690" I.D. .030" wall	
	Water Velocity ft./sec.	Head loss in ft./100 ft.	Water Velocity ft./sec.	Head loss in ft./100 ft.	Water Velocity ft./sec.	Head loss in ft./100 ft.
0.5	.48	.31	.46	.29	.43	.24
1.0	.96	1.05	.92	.95	.86	.76
1.5	1.44	2.11	1.38	1.91	1.29	1.53
2.0	1.92	3.47	1.84	3.14	1.72	2.51
2.5	2.40	5.11	2.30	4.62	2.14	3.68
3.0	2.88	7.02	2.75	6.35	2.57	5.07
3.5	3.36	9.20	3.21	8.32	3.00	6.64
4.0	3.84	11.63	3.67	10.51	3.43	8.40
4.5	4.32	14.30	4.13	12.93	3.86	10.35
5.0	4.80	17.22	4.59	15.56	4.29	12.49
6.0	5.75	23.76	5.51	21.47	5.15	17.21
7.0	6.71	31.22	6.42	28.21	6.00	22.58
8.0	7.67	39.58	7.35	35.75	6.85	28.54
9.0	8.64	48.81	8.25	44.09	7.71	35.35
10.0	9.60	58.90	9.18	53.19	8.57	42.48
11.0	10.60	69.83	10.10	63.06	9.43	50.47
12.0	11.50	81.59	11.00	73.67	10.30	59.10
13.0	12.50	94.18	11.90	85.03	11.20	68.80

*A safety factor of 15-20% should be added to these values.

WATER FRICTION TABLES *(cont.)*

¾" Copper Tubing (S.P.S.)*

Water Flow (gpm)	Type K		Type L		Type M	
	.745" I.D. .065" wall		.785" I.D. .045" wall		.811" I.D. .032" wall	
	Water Velocity ft./sec.	Head loss in ft./100 ft.	Water Velocity ft./sec.	Head loss in ft./100 ft.	Water Velocity ft./sec.	Head loss in ft./100 ft.
1.0	.74	.56	.66	.44	.62	.38
2.0	1.47	1.84	1.33	1.44	1.24	1.23
3.0	2.21	3.73	1.99	2.91	1.86	2.49
4.0	2.94	6.16	2.65	4.81	2.48	4.12
5.0	3.67	9.12	3.31	7.11	3.10	6.09
6.0	4.41	12.57	3.98	9.80	3.72	8.39
7.0	5.14	16.51	4.64	12.86	4.34	11.01
8.0	5.88	20.91	5.30	16.28	4.96	13.94
9.0	6.61	25.77	5.96	20.06	5.59	17.17
10.0	7.35	31.08	6.62	24.19	6.20	20.70
11.0	8.09	36.83	7.29	28.66	6.82	24.52
12.0	8.83	43.01	7.95	33.47	7.44	28.63
13.0	9.56	49.62	8.61	38.61	8.06	33.02
14.0	10.30	56.66	9.27	44.07	8.68	37.69
15.0	11.00	64.11	9.94	49.86	9.30	42.64
16.0	11.80	71.97	10.60	55.97	9.92	47.86
17.0	12.50	80.24	11.25	62.39	10.55	53.35
18.0	13.20	88.92	11.92	69.13	11.17	59.10

*A safety factor of 15-20% should be added to these values.

WATER FRICTION TABLES (cont.)

1" Copper Tubing (S.P.S.)*

Water Flow (gpm)	Type K		Type L		Type M	
	.995" I.D. .065" wall		1.025" I.D. .050" wall		1.055" I.D. .035" wall	
	Water Velocity ft./sec.	Head loss in ft./100 ft.	Water Velocity ft./sec.	Head loss in ft./100 ft.	Water Velocity ft./sec.	Head loss in ft./100 ft.
2.0	.82	.47	.78	.41	.73	.36
3.0	1.24	.95	1.17	.82	1.10	.72
4.0	1.65	1.56	1.56	1.35	1.47	1.18
5.0	2.06	2.30	1.95	2.00	1.83	1.74
6.0	2.48	3.17	2.34	2.75	2.20	2.40
7.0	2.89	4.15	2.72	3.60	2.56	3.14
8.0	3.30	5.25	3.11	4.56	2.93	3.97
9.0	3.71	6.47	3.50	5.61	3.30	4.89
10.0	4.12	7.79	3.89	6.76	3.66	5.89
12.0	4.95	10.76	4.67	9.33	4.40	8.13
14.0	5.77	14.15	5.45	12.27	5.13	10.69
16.0	6.60	17.94	6.22	15.56	5.86	13.55
18.0	7.42	22.14	7.00	19.20	6.60	16.72
20.0	8.24	26.73	7.78	23.18	7.33	20.18
25.0	10.30	39.87	9.74	34.56	9.16	30.09
30.0	12.37	55.33	11.68	47.96	11.00	41.74
35.0	14.42	73.06	13.61	63.31	12.82	55.09
40.0	16.50	93.00	15.55	80.58	14.66	70.11
45.0	18.55	115.10	17.50	99.72	16.50	86.75
50.0	20.60	139.40	19.45	120.70	18.32	105.00

*A safety factor of 15-20% should be added to these values.

WATER FRICTION TABLES (cont.)

1¼" Copper Tubing (S.P.S.)*

Water Flow (gpm)	Type K 1.245" I.D. .065" wall		Type L 1.265" I.D. .055" wall		Type M 1.291" I.D. .042" wall	
	Water Velocity ft./sec.	Head loss in ft./100 ft.	Water Velocity ft./sec.	Head loss in ft./100 ft.	Water Velocity ft./sec.	Head loss in ft./100 ft.
5.0	1.31	.79	1.28	.74	1.22	.67
6.0	1.58	1.09	1.53	1.01	1.47	.92
7.0	1.84	1.43	1.79	1.32	1.71	1.20
8.0	2.11	1.81	2.04	1.67	1.96	1.52
9.0	2.37	2.22	2.30	2.06	2.20	1.87
10.0	2.63	2.67	2.55	2.48	2.45	2.25
12.0	3.16	3.69	3.06	3.42	2.93	3.10
15.0	3.95	5.47	3.83	5.07	3.66	4.60
20.0	5.26	9.13	5.10	8.46	4.89	7.67
25.0	6.58	13.59	6.38	12.59	6.11	11.42
30.0	7.90	18.83	7.65	17.44	7.33	15.82
35.0	9.21	24.83	8.94	23.00	8.55	20.86
40.0	10.50	31.57	10.20	29.24	9.77	26.51
45.0	11.80	38.03	11.50	36.15	11.00	32.77
50.0	13.20	47.20	12.80	43.71	12.20	39.63
60.0	15.80	65.65	15.30	60.78	14.70	55.10
70.0	18.40	86.82	17.90	80.38	17.10	72.86
80.0	21.10	110.70	20.40	102.50	19.60	92.85
90.0	23.70	137.20	23.00	127.00	22.00	115.10
100.0	26.30	166.30	25.50	153.90	24.40	139.40

*A safety factor of 15-20% should be added to these values.

WATER FRICTION TABLES *(cont.)*

1½" Copper Tubing (S.P.S.)*

Water Flow (gpm)	Type K		Type L		Type M	
	1.481" I.D. .072" wall		1.505" I.D. .060" wall		1.527" I.D. .049" wall	
	Water Velocity ft./sec.	Head loss in ft./100 ft.	Water Velocity ft./sec.	Head loss in ft./100 ft.	Water Velocity ft./sec.	Head loss in ft./100 ft.
8.0	1.49	.79	1.44	.73	1.40	.68
9.0	1.67	.97	1.62	.90	1.57	.84
10.0	1.86	1.17	1.80	1.08	1.75	1.01
12.0	2.23	1.61	2.16	1.49	2.10	1.39
15.0	2.79	2.39	2.70	2.21	2.63	2.07
20.0	3.72	3.98	3.60	3.68	3.50	3.44
25.0	4.65	5.91	4.51	5.48	4.38	5.11
30.0	5.58	8.19	5.41	7.58	5.25	7.07
35.0	6.51	10.79	6.31	9.99	6.13	9.31
40.0	7.44	13.70	7.21	12.68	7.00	11.83
45.0	8.37	16.93	8.11	15.67	7.88	14.61
50.0	9.30	20.46	9.01	18.94	8.76	17.66
60.0	11.20	28.42	10.80	26.30	10.50	24.53
70.0	13.00	37.55	12.60	34.74	12.30	32.40
80.0	14.90	47.82	14.40	44.24	14.00	41.25
90.0	16.70	59.21	16.20	54.78	15.80	51.07
100.0	18.60	71.70	18.00	66.34	17.50	61.84
110.0	20.50	85.29	19.80	78.90	19.30	73.55
120.0	22.30	99.95	21.60	92.46	21.00	86.18
130.0	24.20	115.70	23.40	107.00	22.80	99.73

*A safety factor of 15-20% should be added to these values.

WATER FRICTION TABLES (cont.)

2" Copper Tubing (S.P.S.)*

Water Flow (gpm)	Type K		Type L		Type M	
	1.959" I.D. .083" wall		1.985" I.D. .070" wall		2.009" I.D. .058" wall	
	Water Velocity ft./sec.	Head loss in ft./100 ft.	Water Velocity ft./sec.	Head loss in ft./100 ft.	Water Velocity ft./sec.	Head loss in ft./100 ft.
10.0	1.07	.31	1.04	.29	1.01	.27
12.0	1.28	.43	1.24	.40	1.21	.38
14.0	1.49	.56	1.45	.52	1.42	.50
16.0	1.70	.71	1.66	.66	1.62	.63
18.0	1.92	.87	1.87	.82	1.82	.77
20.0	2.13	1.05	2.07	.98	2.02	.93
25.0	2.66	1.55	2.59	1.46	2.53	1.38
30.0	3.19	2.15	3.11	2.01	3.03	1.90
35.0	3.73	2.82	3.62	2.65	3.54	2.50
40.0	4.26	3.58	4.14	3.36	4.05	3.17
45.0	4.79	4.42	4.66	4.15	4.55	3.92
50.0	5.32	5.34	5.17	5.01	5.05	4.73
60.0	6.39	7.40	6.21	6.95	6.06	6.56
70.0	7.45	9.76	7.25	9.16	7.07	8.65
80.0	8.52	12.42	8.28	11.65	8.09	11.00
90.0	9.58	15.36	9.31	14.41	9.10	13.60
100.0	10.65	18.58	10.40	17.43	10.10	16.45
110.0	11.71	22.07	11.40	20.71	11.10	19.55

*A safety factor of 15-20% should be added to these values.

WATER FRICTION TABLES (cont.)

2" Copper Tubing (S.P.S.)*

Water Flow (gpm)	Type K		Type L		Type M	
	1.959" I.D. .083" wall		1.985" I.D. .070" wall		2.009" I.D. .058" wall	
	Water Velocity ft./sec.	Head loss in ft./100 ft.	Water Velocity ft./sec.	Head loss in ft./100 ft.	Water Velocity ft./sec.	Head loss in ft./100 ft.
120.0	12.78	25.84	12.40	24.25	12.10	22.88
130.0	13.85	29.88	13.40	28.04	13.10	26.45
140.0	14.90	34.18	14.50	32.07	14.20	30.26
150.0	16.00	38.75	15.50	36.36	15.20	34.30
160.0	17.00	43.58	16.50	40.89	16.20	38.58
170.0	18.10	48.67	17.60	45.66	17.20	43.08
180.0	19.20	54.01	18.60	50.67	18.20	47.81
190.0	20.20	59.61	19.60	55.92	19.20	52.76
200.0	21.30	65.46	20.70	61.41	20.20	57.94
210.0	22.40	71.57	21.70	67.14	21.20	63.34
220.0	23.40	77.93	22.80	73.10	22.20	68.96
230.0	24.50	84.53	23.80	79.29	23.20	74.80
240.0	25.60	91.38	24.80	85.72	24.30	80.86
250.0	26.60	98.43	25.90	92.37	25.30	87.14
260.0	27.70	105.80	26.90	99.26	26.30	93.63
270.0	28.80	113.40	27.90	106.40	27.30	100.30
280.0	29.80	121.30	29.00	113.70	28.30	107.30
290.0	30.90	129.30	30.00	121.30	29.40	114.40
300.0	32.00	137.60	31.10	129.10	30.40	121.80

*A safety factor of 15-20% should be added to these values.

WATER FRICTION TABLES *(cont.)*

¼" Steel Pipe*

Water Flow (gpm)	Schedule 40 0.364" I.D.			Schedule 80 0.302" I.D.		
	Water Velocity ft./sec.	Water Velocity head ft.	Head loss in ft./100 ft.	Water Velocity ft./sec.	Water Velocity head ft.	Head loss in ft/100 ft.
0.4	1.23	.024	3.70	1.79	.05	9.18
0.6	1.85	.053	7.60	2.69	.11	19.00
0.8	2.47	.095	12.70	3.59	.20	32.30
1.0	3.08	.148	19.10	4.48	.31	48.80
1.2	3.70	.213	26.70	5.38	.45	68.60
1.4	4.32	.290	36.60	6.27	.61	91.70
1.6	4.93	.378	45.60	7.17	.80	118.10
1.8	5.55	.479	56.90	8.07	1.01	147.70
2.0	6.17	.591	69.40	8.96	1.25	180.70
2.4	7.40	.850	98.10	10.75	1.79	256.00
2.8	8.63	1.157	132.00	12.54	2.44	345.00

*A safety factor of 15-20% should be added to these values.

WATER FRICTION TABLES *(cont.)*

⅜" Steel Pipe*

| Water Flow (gpm) | Schedule 40 | | | Schedule 80 | | |
| | 0.493" I.D. | | | 0.423" I.D. | | |
	Water Velocity ft./sec.	Water Velocity head ft.	Head loss in ft./100 ft.	Water Velocity ft./sec.	Water Velocity head ft.	Head loss in ft/100 ft.
0.5	.84	.011	1.26	1.14	.02	2.63
1.0	1.68	.044	4.26	2.28	.08	9.05
1.5	2.52	.099	8.85	3.43	.18	19.00
2.0	3.36	.176	15.00	4.57	.32	32.40
2.5	4.20	.274	22.70	5.71	.51	49.30
3.0	5.04	.395	32.00	6.85	.73	69.60
3.5	5.88	.538	42.70	8.00	.99	93.30
4.0	6.72	.702	55.00	9.14	1.30	120.00
5.0	8.40	1.097	84.20	11.40	2.00	185.00
6.0	10.08	1.580	119.00	13.70	2.90	263.00

*A safety factor of 15-20% should be added to these values.

WATER FRICTION TABLES *(cont.)*

½" Steel Pipe*

| Water Flow (gpm) | Schedule 40 | | | Schedule 80 | | |
| | 0.622" I.D. | | | 0.546" I.D. | | |
	Water Velocity ft./sec.	Water Velocity head ft.	Head loss in ft./100 ft.	Water Velocity ft./sec.	Water Velocity head ft.	Head loss in ft/100 ft.
0.7	.739	.008	.74	.96	.01	1.39
1.0	1.056	.017	1.86	1.37	.03	2.58
1.5	1.580	.039	2.82	2.06	.07	5.34
2.0	2.110	.069	4.73	2.74	.12	9.02
2.5	2.640	.108	7.10	3.43	.18	13.60
3.0	3.170	.156	9.94	4.11	.26	19.10
3.5	3.700	.212	13.20	4.80	.36	25.50
4.0	4.220	.277	17.00	5.48	.47	32.70
4.5	4.750	.351	21.10	6.17	.59	40.90
5.0	5.280	.433	25.80	6.86	.73	50.00
5.5	5.810	.524	30.90	7.54	.88	59.90
6.0	6.340	.624	36.40	8.23	1.05	70.70
6.5	6.860	.732	42.40	8.91	1.23	82.40
7.0	7.390	.849	48.80	9.60	1.43	95.00
7.5	7.920	.975	55.60	10.30	1.60	109.00
8.0	8.450	1.109	63.00	11.00	1.90	123.00
8.5	8.980	1.250	70.70	11.60	2.10	138.00
9.0	9.500	1.400	78.90	12.30	2.40	154.00
9.5	10.030	1.560	87.60	13.00	2.60	171.00
10.0	10.560	1.730	96.60	13.70	2.90	189.00

*A safety factor of 15-20% should be added to these values.

WATER FRICTION TABLES (cont.)

¾" Steel Pipe*

| Water Flow (gpm) | Schedule 40 | | | Schedule 80 | | |
| | 0.824" I.D. | | | 0.742" I.D. | | |
	Water Velocity ft./sec.	Water Velocity head ft.	Head loss in ft./100 ft.	Water Velocity ft./sec.	Water Velocity head ft.	Head loss in ft/100 ft.
1.5	.90	.013	.72	1.11	.02	1.19
2.0	1.20	.023	1.19	1.48	.03	1.99
2.5	1.50	.035	1.78	1.86	.05	2.97
3.0	1.81	.051	2.47	2.23	.08	4.14
3.5	2.11	.069	3.26	2.60	.11	5.48
4.0	2.41	.090	4.16	2.97	.14	7.01
4.5	2.71	.114	5.17	3.34	.17	8.72
5.0	3.01	.141	6.28	3.71	.21	10.60
6.0	3.61	.203	8.80	4.45	.31	14.90
7.0	4.21	.276	11.70	5.20	.42	19.90
8.0	4.81	.360	15.10	5.94	.55	25.60
9.0	5.42	.456	18.80	6.68	.69	32.10
10.0	6.02	.563	23.00	7.42	.86	39.20
11.0	6.62	.681	27.60	8.17	1.04	47.00
12.0	7.22	.722	32.50	8.91	1.23	55.50
13.0	7.82	.951	37.90	9.63	1.44	64.80
14.0	8.42	1.103	43.70	10.40	1.70	74.70
16.0	9.63	1.440	56.40	11.90	2.20	96.70
18.0	10.80	1.820	70.80	13.40	2.80	121.00
20.0	12.00	2.250	86.80	14.80	3.40	149.00

*A safety factor of 15-20% should be added to these values.

WATER FRICTION TABLES *(cont.)*

1" Steel Pipe*

| Water Flow (gpm) | Schedule 40 | | | Schedule 80 | | |
	1.049" I.D.			0.957" I.D.		
	Water Velocity ft./sec.	Water Velocity head ft.	Head loss in ft./100 ft.	Water Velocity ft./sec.	Water Velocity head ft.	Head loss in ft/100 ft.
2.0	.74	.009	.385	.89	.01	.599
3.0	1.11	.019	.787	1.34	.03	1.190
4.0	1.48	.034	1.270	1.79	.05	1.990
5.0	1.86	.054	1.900	2.23	.08	2.990
6.0	2.23	.077	2.650	2.68	.11	4.170
8.0	2.97	.137	4.500	3.57	.20	7.110
10.0	3.71	.214	6.810	4.46	.31	10.800
12.0	4.45	.308	9.580	5.36	.45	15.200
14.0	5.20	.420	12.800	6.25	.61	20.400
16.0	5.94	.548	16.500	7.14	.79	26.300
18.0	6.68	.694	20.600	8.03	1.00	32.900
20.0	7.42	.857	25.200	8.92	1.24	40.300
22.0	8.17	1.036	30.300	9.82	1.50	48.400
24.0	8.91	1.230	35.800	10.70	1.80	57.200
26.0	9.65	1.450	41.700	11.60	2.10	66.800
28.0	10.39	1.680	48.100	12.50	2.40	77.100
30.0	11.10	1.930	55.000	13.40	2.80	88.200
35.0	13.00	2.620	74.100	15.60	3.80	119.000
40.0	14.80	3.430	96.100	17.90	5.00	154.000
45.0	16.70	4.330	121.000	20.10	6.30	194.000

*A safety factor of 15-20% should be added to these values.

WATER FRICTION TABLES (cont.)

1¼" Steel Pipe*

| Water Flow (gpm) | Schedule 40 | | | Schedule 80 | | |
| | 1.380" I.D. | | | 1.278" I.D. | | |
	Water Velocity ft./sec.	Water Velocity head ft.	Head loss in ft./100 ft.	Water Velocity ft./sec.	Water Velocity head ft.	Head loss in ft/100 ft.
4.0	.858	.011	.35	1.00	.015	.51
5.0	1.073	.018	.52	1.25	.024	.75
6.0	1.290	.026	.72	1.50	.034	1.04
7.0	1.500	.035	.95	1.75	.048	1.33
8.0	1.720	.046	1.20	2.00	.062	1.69
10.0	2.150	.072	1.74	2.50	.097	2.55
12.0	2.570	.103	2.45	3.00	.140	3.57
14.0	3.000	.140	3.24	3.50	.190	4.75
16.0	3.430	.183	4.15	4.00	.249	6.10
18.0	3.860	.232	5.17	4.50	.315	7.61
20.0	4.290	.286	6.31	5.00	.388	9.28
25.0	5.360	.431	9.61	6.25	.607	14.20
30.0	6.440	.644	13.60	7.50	.874	20.10
35.0	7.510	.876	18.20	8.75	1.190	27.00
40.0	8.580	1.140	23.50	10.00	1.550	34.90
50.0	10.700	1.790	36.20	12.50	2.430	53.70
60.0	12.900	2.570	51.50	15.00	3.500	76.50
70.0	15.000	3.500	69.50	17.50	4.760	103.00
80.0	17.200	4.530	90.20	20.00	6.210	134.00
90.0	19.300	5.790	114.00	22.50	7.860	168.00

*A safety factor of 15-20% should be added to these values.

WATER FRICTION TABLES (cont.)

1½" Steel Pipe*

Water Flow (gpm)	Schedule 40 1.610" I.D.			Schedule 80 1.500" I.D.		
	Water Velocity ft./sec.	Water Velocity head ft.	Head loss in ft./100 ft.	Water Velocity ft./sec.	Water Velocity head ft.	Head loss in ft/100 ft.
4.0	.63	.006	.166	.73	.01	.233
5.0	.79	.010	.246	.91	.01	.346
6.0	.95	.014	.340	1.09	.02	.478
7.0	1.10	.019	.447	1.27	.03	.630
8.0	1.26	.025	.567	1.45	.03	.800
9.0	1.42	.031	.701	1.63	.04	.990
10.0	1.58	.039	.848	1.82	.05	1.200
12.0	1.89	.056	1.180	2.18	.07	1.610
14.0	2.21	.076	1.510	2.54	.10	2.140
16.0	2.52	.099	1.930	2.90	.13	2.740
18.0	2.84	.125	2.400	3.27	.17	3.410
20.0	3.15	.154	2.920	3.63	.20	4.150
22.0	3.47	.187	3.480	3.99	.25	4.960
24.0	3.78	.222	4.100	4.36	.30	5.840
26.0	4.10	.261	4.760	4.72	.35	6.800
28.0	4.41	.303	5.470	5.08	.40	7.820
30.0	4.73	.347	6.230	5.45	.46	8.910
32.0	5.04	.395	7.040	5.81	.52	10.100
34.0	5.36	.446	7.900	6.17	.59	11.300
36.0	5.67	.500	8.800	6.54	.66	12.600
38.0	5.99	.577	9.760	6.90	.74	14.000
40.0	6.30	.618	10.800	7.26	.82	15.400

*A safety factor of 15-20% should be added to these values.

WATER FRICTION TABLES *(cont.)*

1½" Steel Pipe*

Water Flow (gpm)	Schedule 40			Schedule 80		
	1.610" I.D.			1.500" I.D.		
	Water Velocity ft./sec.	Water Velocity head ft.	Head loss in ft./100 ft.	Water Velocity ft./sec.	Water Velocity head ft.	Head loss in ft/100 ft.
42.0	6.62	.681	11.80	7.63	.90	16.90
44.0	6.93	.747	12.90	7.99	.99	18.50
46.0	7.25	.817	14.00	8.35	1.08	20.10
48.0	7.56	.889	15.20	8.72	1.18	21.80
50.0	7.88	.965	16.50	9.08	1.28	23.60
55.0	8.67	1.170	19.80	9.99	1.55	28.40
60.0	9.46	1.390	23.40	10.90	1.80	33.60
65.0	10.24	1.630	27.30	11.80	2.20	39.20
70.0	11.03	1.890	31.50	12.70	2.50	45.30
75.0	11.80	2.170	36.00	13.60	2.90	51.80
80.0	12.60	2.470	40.80	14.50	3.30	58.70
85.0	13.40	2.790	45.90	15.40	3.70	66.00
90.0	14.20	3.130	51.30	16.30	4.10	73.80
95.0	15.00	3.480	57.00	17.20	4.60	82.00
100.0	15.80	3.860	63.00	18.20	5.10	90.70
110.0	17.30	4.670	75.80	20.00	6.20	109.30
120.0	18.90	5.560	89.90	21.80	7.40	129.60
130.0	20.50	6.520	105.00	23.60	8.70	151.60
140.0	22.10	7.560	122.00	25.40	10.00	175.00
150.0	23.60	8.680	139.00	27.20	11.50	201.00
160.0	25.20	9.880	158.00	29.00	13.10	228.00
170.0	26.80	11.150	178.00	30.90	14.80	257.00
180.0	28.40	12.500	199.00	32.70	16.60	288.00

*A safety factor of 15-20% should be added to these values.

WATER FRICTION TABLES (cont.)

2" Steel Pipe*

Water Flow (gpm)	Schedule 40 2.067" I.D.			Schedule 80 1.939" I.D.		
	Water Velocity ft./sec.	Water Velocity head ft.	Head loss in ft./100 ft.	Water Velocity ft./sec.	Water Velocity head ft.	Head loss in ft/100 ft.
5.0	.478	.004	.074	.54	.00	.101
6.0	.574	.005	.102	.65	.01	.139
7.0	.669	.007	.134	.76	.01	.182
8.0	.765	.009	.170	.87	.01	.231
9.0	.860	.012	.209	.98	.01	.285
10.0	.956	.014	.252	1.09	.02	.343
12.0	1.150	.021	.349	1.30	.03	.476
14.0	1.340	.028	.461	1.52	.04	.629
16.0	1.530	.036	.586	1.74	.05	.800
18.0	1.720	.046	.725	1.96	.06	.991
20.0	1.910	.057	.878	2.17	.07	1.160
22.0	2.100	.069	1.050	2.39	.09	1.380
24.0	2.290	.082	1.180	2.61	.11	1.620
26.0	2.490	.096	1.370	2.83	.12	1.880
28.0	2.680	.111	1.570	3.04	.14	2.160
30.0	2.870	.128	1.820	3.26	.17	2.460
35.0	3.350	.174	2.380	3.80	.22	3.280
40.0	3.820	.227	3.060	4.35	.29	4.210
45.0	4.300	.288	3.820	4.89	.37	5.260
50.0	4.780	.355	4.660	5.43	.46	6.420
55.0	5.260	.430	5.580	5.98	.56	7.700
60.0	5.740	.511	6.580	6.52	.66	9.090

*A safety factor of 15-20% should be added to these values.

WATER FRICTION TABLES (cont.)

2" Steel Pipe*

Water Flow (gpm)	Schedule 40			Schedule 80		
	2.067" I.D.			1.939" I.D.		
	Water Velocity ft./sec.	Water Velocity head ft.	Head loss in ft./100 ft.	Water Velocity ft./sec.	Water Velocity head ft.	Head loss in ft/100 ft.
65.0	6.21	.600	7.66	7.06	.77	10.59
70.0	6.69	.696	8.82	7.61	.90	12.20
75.0	7.17	.799	10.10	8.15	1.03	13.90
80.0	7.65	.909	11.40	8.69	1.17	15.80
85.0	8.13	1.030	12.80	9.03	1.27	17.70
90.0	8.60	1.150	14.30	9.78	1.49	19.80
95.0	9.08	1.280	15.90	10.30	1.60	22.00
100.0	9.56	1.420	17.50	10.90	1.80	24.30
110.0	10.52	1.720	21.00	12.00	2.20	29.20
120.0	11.50	2.050	24.90	13.00	2.60	34.50
130.0	12.40	2.400	29.10	14.10	3.10	40.30
140.0	13.40	2.780	33.60	15.20	3.60	46.60
150.0	14.30	3.200	38.40	16.30	4.10	53.30
160.0	15.30	3.640	43.50	17.40	4.70	60.50
170.0	16.30	4.110	49.00	18.50	5.30	68.10
180.0	17.20	4.600	54.80	19.60	6.00	76.10
190.0	18.20	5.130	60.90	20.60	6.60	84.60
200.0	19.10	5.680	67.30	21.70	7.30	93.60
220.0	21.00	6.880	81.10	23.90	8.90	113.00
240.0	22.90	8.180	96.20	26.90	10.60	134.00
260.0	24.90	9.600	113.00	28.30	12.40	157.00
280.0	26.80	11.140	130.00	30.40	14.40	181.00
300.0	28.70	12.800	149.00	32.60	16.50	208.00

*A safety factor of 15-20% should be added to these values.

WEIR DISCHARGE VOLUME RATE IN GPM

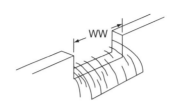

| Head | Weir Width (WW) in Feet | | | GPM per foot over 5 |
| (in.) | 1 | 3 | 5 | |
	Gallons per Minute			
1	35	107	179	36
1½	64	197	329	66
2	98	302	506	102
2½	136	421	705	142
3	178	552	926	187
4	269	845	1420	288
5	369	1174	1978	402
6	476	1534	2592	529
7	–	1922	3255	667
8	–	2335	3963	814
9	–	2769	4713	972
10	–	3225	5501	1138
12	–	4189	7181	1496

NOZZLE DISCHARGE RATES IN GPM

Nozzle Pressure (psi)	Diameter of Nozzle Orifice (in.)																
	1/16	1/8	3/16	1/4	5/16	3/8	7/16	1/2	9/16	5/8	3/4	7/8	1	1 1/8
	Gallons per Minute													
10	.38	1.48	3.3	5.9	9.24	13.3	18.1	23.6	30.2	36.9	53.3	72.5	94.8	120
15	.45	1.81	4.1	7.2	11.4	16.3	22.4	28.9	36.7	45.2	65.1	88.7	116	147
20	.53	2.09	4.7	8.3	13.1	18.7	25.6	33.4	42.4	52.2	75.4	102	134	169
25	.59	2.34	5.3	9.3	14.6	21.0	28.7	37.3	47.3	58.2	84.0	115	149	189
30	.64	2.56	5.8	10.2	16.0	23.1	31.4	40.9	51.9	63.9	92.2	126	164	208
35	.69	2.78	6.2	11.1	17.1	25.0	33.8	44.2	56.1	69.0	99.8	136	177	224
40	.74	2.96	6.7	11.7	18.4	26.6	36.2	47.3	59.9	73.8	106	145	189	239
45	.79	3.14	7.1	12.6	19.5	28.2	38.3	50.1	63.4	78.2	113	153	200	254
50	.83	3.30	7.4	13.2	20.6	29.9	40.5	52.8	67.0	82.5	119	162	211	268
60	.90	3.62	8.2	14.5	22.6	32.6	44.3	57.9	73.3	90.4	130	177	232	293
70	.98	3.91	8.8	15.7	24.4	35.3	47.9	62.6	79.3	97.8	141	192	251	317
80	1.05	4.19	9.4	16.8	26.1	37.6	51.2	66.8	84.8	105	151	205	268	339
90	1.11	4.43	10.0	17.7	27.8	40.1	54.5	70.8	90.3	111	160	218	285	360
100	1.17	4.67	10.4	18.7	29.2	42.2	57.3	74.9	95.0	117	169	229	300	379
120	1.23	5.17	11.5	20.4	31.8	46.0	62.4	81.8	103	128	184	250	327	413
140	1.28	5.70	12.4	22.1	34.4	49.8	67.6	88.3	112	138	199	271	354	447
160	1.32	6.30	13.3	23.6	36.9	53.3	72.3	94.6	120	148	213	289	378	478
180	1.36	6.92	14.1	25.0	39.0	56.4	76.5	100	127	156	225	306	400	506
200	1.38	7.52	14.9	26.4	41.1	59.5	81.6	106	134	165	238	323	423	535

Note: These rates are only theoretical. Actual values will be 95% of the above depending on nozzle condition.

HORIZONTAL PIPE DISCHARGE RATE IN GPM

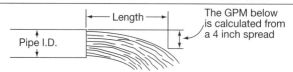

Length — Pipe I.D. — The GPM below is calculated from a 4 inch spread

Length	Gallons per minute per Pipe I.D. (in.)											
(in.)	1	1¼	1½	2	2½	3	4	5	6	8	10	12
4	6	10	13	22	31	48	83	–	–	–	–	–
5	7	12	17	27	39	61	104	163	–	–	–	–
6	8	15	20	33	47	73	125	195	285	–	–	–
7	10	17	23	38	55	85	146	228	334	580	–	–
8	11	20	26	44	62	97	166	260	380	665	1060	–
9	13	22	30	49	70	110	187	293	430	750	1190	1660
10	14	24	33	55	78	122	208	326	476	830	1330	1850
11	16	27	36	60	86	134	229	360	525	915	1460	2020
12	17	29	40	66	94	146	250	390	570	1000	1600	2220
13	18	31	43	71	102	158	270	425	620	1080	1730	2400
14	20	34	46	77	109	170	292	456	670	1160	1860	2590
15	21	36	50	82	117	183	312	490	710	1250	2000	2780
16	23	39	53	88	125	196	334	520	760	1330	2120	2960
17	–	41	56	93	133	207	355	550	810	1410	2260	3140
18	–	–	60	99	144	220	375	590	860	1500	2390	3330
19	–	–	–	110	148	232	395	620	910	1580	2520	3500
20	–	–	–	–	156	244	415	650	950	1660	2660	3700
21	–	–	–	–	–	256	435	685	1000	1750	2800	3890
22	–	–	–	–	–	–	460	720	1050	1830	2920	4060
23	–	–	–	–	–	–	–	750	1100	1910	3060	4250
24	–	–	–	–	–	–	–	–	1140	2000	3200	4440

VERTICAL PIPE DISCHARGE RATE IN GPM

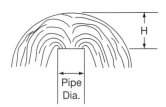

Pipe Dia.

The formula below is an approximation of a typical vertical pipe output.

GPM= $\sqrt{H}$ x K x (Dia.)^2 x 5.68

where:

 GPM = gallons per minute
 H = height (in.)
 Dia. = diameter of pipe (in.)
 K = constant from 0.87 to 0.97 for pipe diameters from
  2 to 6 inches and heights up to 24 inches.

OAKUM AND LEAD REQUIREMENTS FOR CAULKED JOINTS PER N.P.S.

Material	2"	3"	4"	5"	6"	7"	8"	10"
Oakum (feet)	3	4½	5	6½	7½	8½	9½	12
Lead (lbs.)	1½	2¼	3	3¾	4½	5¼	6	7½

LENGTH OF WIPED JOINTS				
Type of Pipe and Nominal Pipe Size (in.)	**One-hand system**		**Two-hand system**	
	Length of joint (in.)	Size of cloth (in.)	Length of joint (in.)	Size of cloth (in.)
½ water	2	3 X 3	2¼	3 X 4
¾ water	2	3 X 3	2⅜	3 X 4
1 water	2	3 X 3	2⅜	3 X 4
1¼ water	2	3 X 3	2½	3¼ X 4
1¼ waste	2	3 X 3	2⅜	3 X 4
1½ water	2	3 X 3	2½	3¼ X 4
1½ waste	2	3 X 3	2⅜	3 X 4
2 waste	2	3 X 3	2⅜	3¼ X 4
3 waste	2	3 X 3	2½	3¼ X 4
4 waste	1¾	3 x 3, 6 x 6	2¾	3¼ x 4, 3¼ X 5
2 vertical	1¾	3 X 3	2	3 X 2½
3 vertical	1¾	3 X 3	2	3 X 2½
4 vertical	1¾	3 X 3	2	3 X 2½

FLANGE UNION GASKETS

Apply graphite to one side of gasket to make removal easier.
Suitable gasket material should be made of the following:

Cold-water piping	Sheet rubber or asbestos sheet packing (if permitted)
Hot-water piping	Rubber or asbestos (if permitted)
Gas piping	Leather or asbestos (if permitted)
Oil lines	Metallic or asbestos (if permitted)
Gasoline conduction	Metallic

HANGER ROD SIZES

Iron Pipe Size (in.)	Rod Size (in.)	Iron Pipe Size (in.)	Rod Size (in.)
⅛ to ½	¼	6	¾
¾ to 2	⅜	8 to 12	⅞
2½ to 3	½	14 to 16	1
4 to 5	⅝		

MELTING POINTS (F)

50-50 solder melts at 362°   **Zinc** melts at 790°

Tin melts at 449°   **Steel** melts at 2400°-2700°

Lead melts at 622°   **Pure iron** melts at 2730°

MINIMUM GRADE FALL

The absolute minimum fall or grade for foundation or subsoil
drainage lines is 1 inch for every 20 feet of length.

CHAPTER 4
Drainage Systems

DRAINAGE SYSTEM MATERIALS AND USAGE			
Sanitary Drainage and Waste Systems			
Pipe Material	**Building Sewer**	**Underground Within Buildings**	**Aboveground Within Buildings**
Asbestos-cement sewer	X	–	–
ABS Schedule 40	X	X	X
Borosilicate glass	–	–	X
Brass	–	–	X
Cast iron soil	X	X	X
Concrete drain	X	–	–
Copper, DWV	X	X	X
Copper, Type K, L, M	X	X	X
Galvanized steel	–	–	X
Lead	–	X	X
PVC Schedule 40	X	X	X
Vitrified clay, standard	X	–	–
Vitrified clay, extra strength	X	X	–
Wrought iron	–	–	X

DRAINAGE SYSTEM MATERIALS AND USAGE (*cont.*)

Vent Pipes for Drainage and Waste Systems

Pipe Material	Building Sewer	Underground Within Buildings	Aboveground Within Buildings
ABS Schedule 40	N/A	X	X
Borosilicate glass	N/A	–	X
Brass	N/A	–	X
Cast iron soil	N/A	X	X
Copper, DWV	N/A	X	X
Copper, Type K, L, M	N/A	X	X
Galvanized steel	N/A	–	X
Lead	N/A	–	X
PVC Schedule 40	N/A	X	X
Vitrified clay, extra strength	N/A	X	–
Wrought iron	N/A	–	X

DRAINAGE SYSTEM MATERIALS AND USAGE (*cont.*)

Storm Drainage Systems

Pipe Material	Building Sewer	Underground Within Buildings	Aboveground Within Buildings
Asbestos-cement pipe	X	–	–
ABS Schedule 40	X	X	X
Brass	–	–	X
Cast iron soil	X	X	X
Concrete drain	X	–	–
Copper DWV	X	X	X
Copper, Type K, L, M	X	X	X
Galvanized steel	–	–	X
Lead	–	–	X
PVC Schedule 40	X	X	X
Vitrified clay, extra strength	X	X	–
Wrought iron	–	–	X

DRAINAGE SYSTEM MATERIALS AND USAGE (*cont.*)

Chemical Waste and Acid Systems

Pipe Material	Building Sewer	Underground Within Buildings	Aboveground Within Buildings
ABS Schedule 40	N/A	X	X
Borosilicate glass	N/A	–	X
High silicon content iron	N/A	X	X
Lead	N/A	–	X
Plastic lined	N/A	X	X
PVC Schedule 40	N/A	X	X
Vitrified clay, extra strength	N/A	X	–

TYPICAL SEWAGE FLOW

Location	Gallons per Day
Airports	15 per employee
	5 per person
Assembly halls	2 per seat
Bowling alleys	75 per lane
Boarding houses	50 per person
Churches	
(Small)	5 per seat
(Small with kitchen)	7 per seat
(Large)	7 per seat
(Large with kitchen)	8 per seat
Dance halls	2 per person
Day camps	15 per person
Day camps (with meals)	25 per person
Factories (per shift, exclusive of industrial wastes)	
No showers	25 per employee
With showers	35 per employee
Homes	
Subdivisions	75 per person
Individual residences	100 per person
Luxury residences	150 per person
Hospitals	150 per bed

TYPICAL SEWAGE FLOW *(cont.)*	
Location	**Gallons per Day**
Hotels	
with connecting baths	50 per bed
with private bath and no	60 per bed
kitchen (2 guests per room)	
with private bath and	100 per bed
kitchen (2 guests per room)	
Kitchen wastes from hotels, camps, boarding houses, etc. that serve meals daily	10 per person
Laundries (coin-operated)	400 per machine
Marinas (note: rate is per hour)	
Flush toilets	36 per fixture per hour
Urinals	10 per fixture per hour
Wash basins	15 per fixture per hour
Showers	150 per fixture per hour
Mobile homes	250 per space
Motels	
No kitchen	50 per bed
With kitchen	60 per bed
Nursing homes	125 per person
Offices	20 per employee
Public institutions (other than hospitals)	100 per person

TYPICAL SEWAGE FLOW *(cont.)*

Location	Gallons per Day

Public parks
  Toilet wastes only
  Bathhouse with showers
    and toilets
	
	5 per person
	
	10 per person

Restaurants
  (toilet and kitchen wastes
  per serving capacity)
	25 per person

Schools
  Elementary
    (toilets and lavatories only)
  Intermediate and high
    (with cafeteria)
    (with cafeteria and showers)
  Office staff and teachers
	
	15 per student
	
	25 per student
	35 per student
	20 per person

Service stations
  No bays
  With one bay
  Each additional bay
	
	10 per vehicle served
	1000 per day
	500 per day

Stores (retail)	400 per bathroom with toilet
Swimming pools	10 per person

Theaters:
  Auditorium
  Drive-in
	
	5 per seat
	10 per car

Trailer parks	50 per person
Work or construction camps	50 per person

SOIL ABSORPTION CAPACITY/LEACHING AREA

Type of Soil	Required Square Feet of Leaching Area per 100 Gallons	Maximum Absorption Capacity of Leaching Area per Day in Gal. per Sq. Ft.
Coarse sand or gravel	20	5.00
Fine sand	25	4.00
Sandy loam or sandy clay	40	2.50
Clay with sand or gravel	90	1.10
Clay with small amount of sand or gravel	120	.83

SEPTIC TANK CAPACITY TO LEACHING AREA

Required Square Feet of Leaching Area per 100 Gallons of Septic Tank Capacity	Allowable Maximum Septic Tank Capacity in Gallons
20 to 25	7,500
40	5,000
60	3,500
90	3,000

SEPTIC TANKS, SINGLE-FAMILY RESIDENCES

Single Family Residence Number of Bedrooms	Minimum Septic Tank Capacity (gal.)
1 to 2	750
3	1,000
4	1,200
5 to 6	1,500

Add 150 gallons per additional bedroom. Septic tank sizes include sludge storage capacity and the connection of food waste disposal units.

SEPTIC TANKS, MULTI-FAMILY RESIDENCES

Number of Dwelling Units One Bedroom per Unit	Minimum Septic Tank Capacity (gal.)
2	1,200
3	1,500
4	2,000
5	2,250
6	2,500
7	2,750
8	3,000
9	3,250
10	3,500

Add 150 gallons for each additional bedroom in a unit.
Add 250 gallons for each dwelling unit over 10.
Septic tank sizes include sludge storage capacity and the connection of food waste disposal units.

CAPACITY OF SEPTIC TANK PER TOTAL FIXTURES

Single-Family Residences Number of Bedrooms	Maximum Fixture Units Served	Minimum Septic Tank Capacity (gal.)
1 to 2	15	750
3	20	1000
4	25	1200
5 to 6	33	1500

DRAINFIELD CONVERSION

Number of Bedrooms	Drainfield Required (Square feet)	Conventional Block Drain (Linear Feet)	Corrugated 4" Plastic Tubing (Linear Feet)
2	100	25	40
3	125	32	50
4	150	38	60
5	175	44	70

SEPTIC TANK DIAGRAM

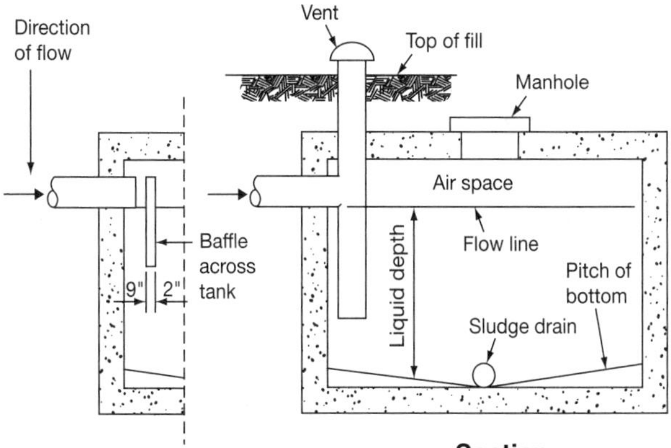

Direction of flow

Baffle across tank

9" 2"

Alternate
with baffle on inlet

Vent

Top of fill

Manhole

Air space

Liquid depth

Flow line

Pitch of bottom

Sludge drain

Section

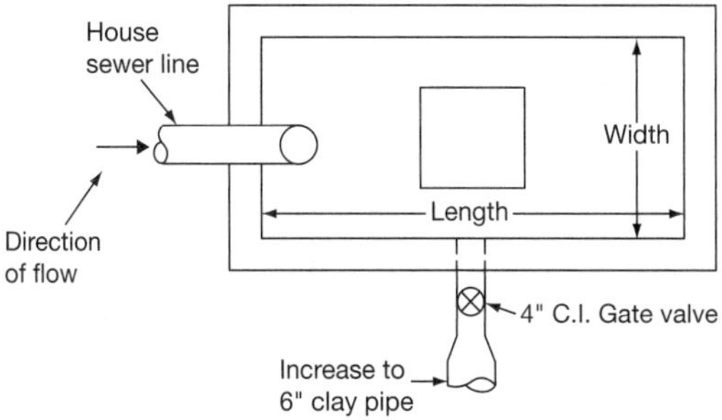

House sewer line

Width

Length

Direction of flow

4" C.I. Gate valve

Increase to 6" clay pipe

Plan

4-11

DISTRIBUTION BOX DIAGRAM

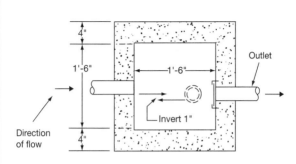

4"
1'-6"
1'-6"
Invert 1"
Outlet
4"
Direction of flow

Section

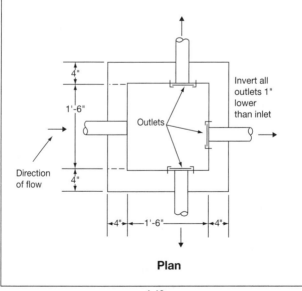

4"
1'-6"
4"
Outlets
Invert all outlets 1" lower than inlet
Direction of flow
4" — 1'-6" — 4"

Plan

4-12

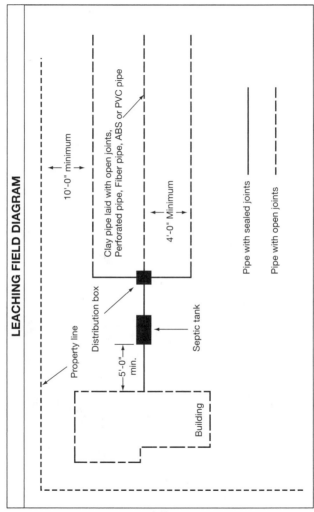

LEACHING FIELD DIAGRAM

Property line

10'-0" minimum

Clay pipe laid with open joints,
Perforated pipe, Fiber pipe, ABS or PVC pipe

4'-0" Minimum

Distribution box

Septic tank

5'-0" min.

Building

Pipe with sealed joints —————

Pipe with open joints — — — —

4-13

12 INCH STONE LEACHING CESSPOOL DIAGRAM

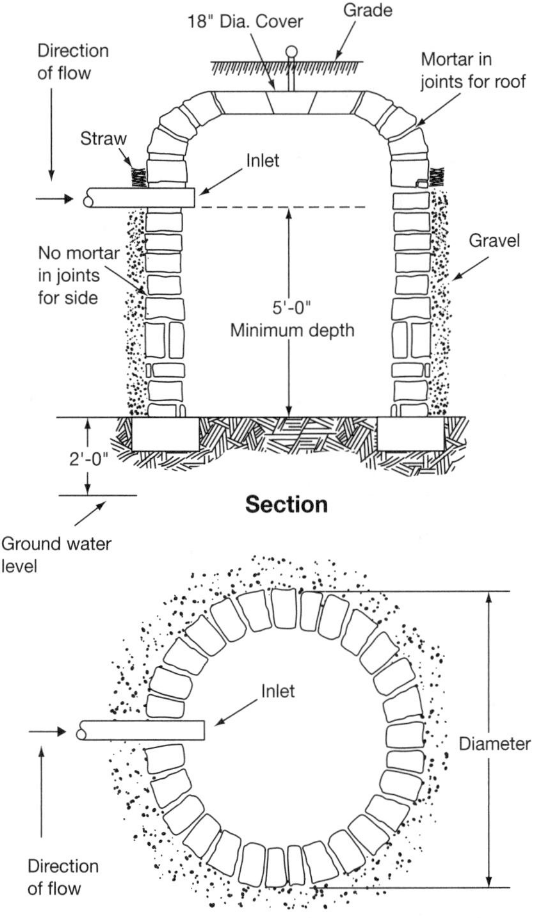

Section

Labels in Section:
- 18" Dia. Cover
- Grade
- Direction of flow
- Mortar in joints for roof
- Straw
- Inlet
- No mortar in joints for side
- Gravel
- 5'-0" Minimum depth
- 2'-0"
- Ground water level

Plan

Labels in Plan:
- Inlet
- Diameter
- Direction of flow

4-14

PUMP TANK DIAGRAM FOR A SAND MOUND

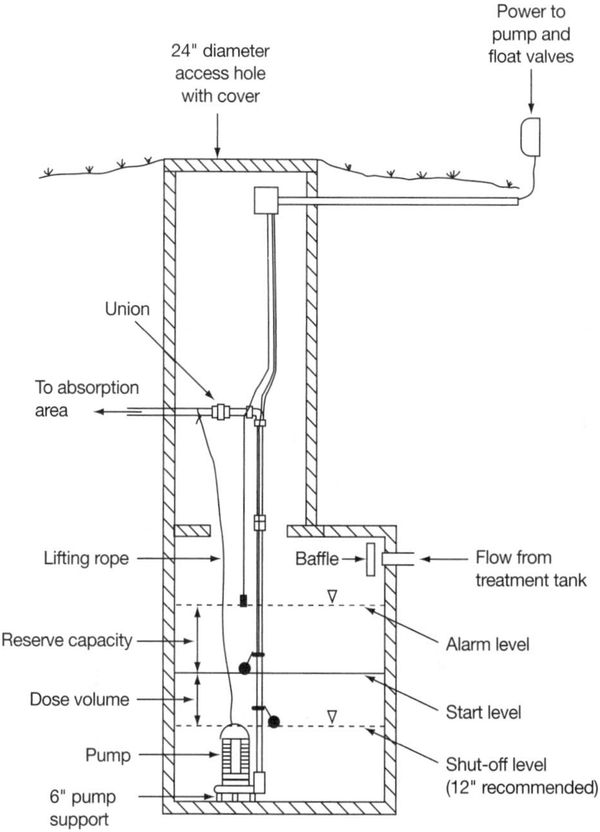

Power to pump and float valves

24" diameter access hole with cover

Union

To absorption area ←

Lifting rope

Baffle → ← Flow from treatment tank

Reserve capacity

Alarm level

Dose volume

Start level

Pump

Shut-off level (12" recommended)

6" pump support

SAND MOUND DIAGRAM

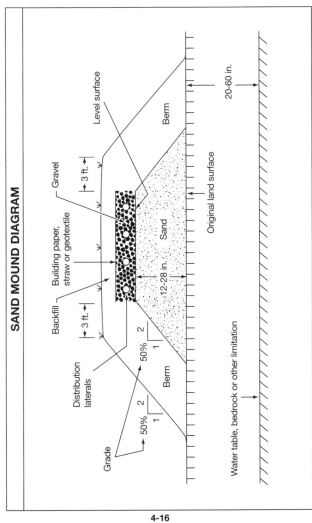

Level surface

Gravel

Building paper, straw or geotextile

Backfill

3 ft.

3 ft.

Distribution laterals

Grade

50% $\frac{2}{1}$

50% $\frac{2}{1}$

Berm

Berm

Sand

12–28 in.

Original land surface

20–60 in.

Water table, bedrock or other limitation

4-16

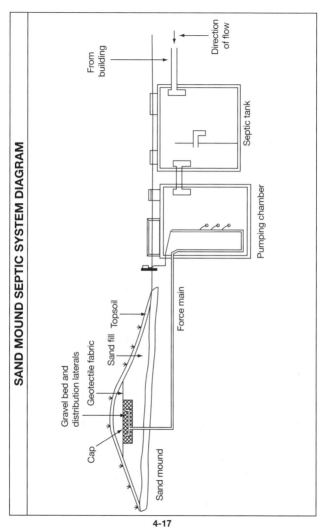

SAND MOUND SEPTIC SYSTEM DIAGRAM

From building

Direction of flow

Septic tank

Pumping chamber

Force main

Gravel bed and distribution laterals

Geotectile fabric

Sand fill   Topsoil

Cap

Sand mound

4-17

LOCATION OF SEWAGE DISPOSAL SYSTEM MEASURED IN FEET

Minimum Horizontal Distance From:	Building Sewer	Septic Tank	Disposal Field	Seepage Pit or Cesspool
Building	2	5	8	8
Private property	Clear	5	5	8
Water supply well	50	50	100	150
Water service line	1	5	5	5
Public water main	10	10	10	10
Distribution box –	–	5	5	
Disposal field	–	5	4	5
Seepage pit or cesspool	–	5	5	2
Trees	–	10	–	10
Streams	50	50	50	100

All trenches running parallel and deeper than the footing of a building must be at least 45° from the footing.

DRAIN AND TRAP SIZES FOR FIXTURE UNITS

Fixture	Number of Fixture Units	Drain and Trap Size (in.)
Lavatory (wash basin)	1	1¼
Lavatory (inset)	2	1½
Water closet, flush tank	4	3
Water closet, flushomatic	6	3
Bathtub	2	1½
Shower (single stall)	2	2
Kitchen sink	2	1½
Dishwasher	2	1½
Laundry tub	2	1½
Washing Machine	2	2
Service Sink	3	2
Drinking fountain	1	1¼
Urinal (stall)	2	2
Bidet	2	1½
Floor drain	2	2

MAXIMUM NUMBER OF FIXTURE UNITS ALLOWED FOR CONNECTION TO ANY PART OF THE BUILDING DRAIN OR SEWER PER THE GIVEN FALL

Drain or Sewer Pipe Size (in.)	1/16" Fall	1/8" Fall	1/4" Fall	1/2" Fall
2	–	–	21	26
2½	–	–	24	31
3	–	20	27	36
4	–	180	216	250
5	–	390	480	575
6	–	700	840	1,000
8	1,400	1,600	1,920	2,300
10	2,500	2,900	3,500	4,200
12	3,900	4,600	5,600	6,700

MAXIMUM UNIT LOADING AND LENGTH OF DRAINAGE AND VENT PIPING

Drain or Vent Pipe Size (in.)	Maximum units drainage piping		Maximum length (ft.) drainage piping		Vertical and horizontal vent piping	
	Vertical	Horizontal	Vertical	Horizontal	Maximum units	Maximum length (ft.)
1¼	1	1	45	Unlimited	1	45'
1½	2	1	65	Unlimited	8	60'
2	16	8	85	Unlimited	24	120'
2½	32	14	148	Unlimited	48	180'
3	48	35	212	Unlimited	84	212'
4	256	216	300	Unlimited	256	300'
5	600	428	390	Unlimited	600	390'
6	1380	720	510	Unlimited	1380	510'
8	3600	2640	750	Unlimited	3600	750'
10	5600	4680	–	Unlimited	–	–
12	8400	8200	–	Unlimited	–	–

FLOW AND PIPE SIZE PER NUMBER OF PLUMBING FIXTURES

Number of Fixtures	Bathtubs		Lavatories and Sinks		Service Sinks	
	Pipe Size (in.)	Faucet (gpm)	Pipe Size (in.)	Faucet (gpm)	Pipe Size (in.)	Faucet (gpm)
1	¾	15	½	4	¾	15
2	1	30	½	8	1	25
4	1¼	40	¾	12	1¼	40
8	1½	80	1	24	1½	64
12	2	96	1	30	1½	84
16	2	112	1¼	40	2	96
24	3	144	1¼	48	2	120
32	3	192	1½	64	2	150
40	3	240	1½	75	3	200

FLOW AND PIPE SIZE PER NUMBER OF PLUMBING FIXTURES (cont.)

Total Number of Fixtures	Urinals				Water Closets		
	Pipe Size (in.)	Tank (gpm)	Pipe Size (in.)	Flusho-meter (gpm)	Pipe Size (in.)	Tank (gpm)	Flusho-meter (gpm)
1	½	6	1	25	½	8	30
2	¾	12	1¼	37	¾	16	50
4	1	20	1¼	45	1	24	80
8	1¼	32	1½	75	1¼	48	120
12	1¼	43	1½	85	1½	60	140
16	1¼	56	2	100	1½	80	160
24	1½	72	2	125	2	96	200
32	2	90	2	150	2	128	250
40	2	120	2	175	2	150	300

SIZING HORIZONTAL STORM DRAINS

Drain Size (in.)	Maximum Roof Area in Square Feet For Drains at Various Slopes		
	1/8" Slope	1/4" Slope	1/2" Slope
3	822	1,160	1,644
4	1,880	2,650	3,760
5	3,340	4,720	6,680
6	5,350	7,550	10,700
8	11,500	16,300	23,000
10	20,700	29,200	41,400
12	33,300	47,000	66,600
15	59,500	84,000	119,000

SIZING VERTICAL RAINWATER LEADERS

Leader Size (in.)	Maximum Roof Area (Sq. Ft.)
2	720
2½	1,300
3	2,200
4	4,600
5	8,650
6	13,500
8	29,000

Values are based on a maximum rate of 4 inches of rainfall per hour.

CHAPTER 5
Pumps

NET POSITIVE SUCTION HEAD

Net Positive Suction Head is the total suction head measured in feet of liquid from which is deducted the absolute vapor pressure (also measured in feet of liquid) of the liquid being pumped. There are two formulas:

Suction lift or the supply level being below the pump centerline.

$$NPSH = \left(h_a - h_{vpa}\right) - \left(h_{st} - h_{fs}\right)$$

Flooded suction or the supply is above the pump centerline.

$$NPSH = \left(h_a - h_{vpa}\right) + \left(h_{st} - h_{fs}\right)$$

where:

h_a = absolute pressure (in feet of liquid) on the surface of the liquid supply level. (This is the atmospheric pressure for an open tank or sump or the absolute pressure existing in a closed tank.)

h_{vpa} = head (in feet) corresponding to the vapor pressure of the liquid at the temperature it is being pumped.

h_{st} = static height (in feet) that the liquid supply level is above or below the pump centerline.

h_{fs} = All suction line head friction losses (in feet) including entrance losses and friction losses through the pipe, fittings and valves.

WATER PRESSURE IN POUNDS PER SQUARE INCH
WITH EQUIVALENT FEET HEADS

PSI	Feet Head	PSI	Feet Head
1	2.31	100	230.90
2	4.62	110	253.98
3	6.93	120	277.07
4	9.24	130	300.16
5	11.54	140	323.25
6	13.85	150	346.34
7	16.16	160	369.43
8	18.47	170	392.52
9	20.78	180	415.61
10	23.09	200	461.78
15	34.63	250	577.24
20	46.18	300	692.69
25	57.72	350	808.13
30	69.27	400	922.58
40	92.36	500	1154.48
50	115.45	600	1385.39
60	138.54	700	1616.30
70	161.63	800	1847.20
80	184.72	900	2078.10
90	207.81	1000	2309.00

WATER FEET HEAD TO POUNDS PER SQUARE INCH

Feet Head	PSI	Feet Head	PSI
1	.43	50	21.65
2	.87	60	25.99
3	1.30	70	30.32
4	1.73	80	34.65
5	2.17	90	38.98
6	2.60	100	43.34
7	3.03	110	47.64
8	3.46	120	51.97
9	3.90	130	56.30
10	4.33	140	60.63
15	6.50	150	64.96
20	8.66	160	69.29
25	10.83	170	73.63
30	12.99	180	77.96
40	17.32	200	86.62

AIR PRESSURE IN POUNDS PER SQUARE INCH

Altitude In Feet	PSI
Sea Level	1.6
10,000 ft.	2.7
20,000 ft.	4.3
30,000 ft.	6.4
40,000 ft.	10.2
50,000 ft.	14.7

GENERAL PUMP FORMULAS

$$\text{Pressure (psi)} = \frac{\left(\text{Head (ft.)} \times \text{Specific Gravity}\right)}{2.31}$$

$$\text{Head (ft.)} = \frac{\left(\text{Pressure (psi)} \times 2.31\right)}{\text{Specific Gravity}}$$

$$\text{Pipe velocity (ft. per second)} = \frac{.408 \times \text{GPM}}{\left(\text{pipe diameter}\right)^2}$$

$$\text{Velocity head (ft.)} = \frac{\left(\text{pipe velocity ft. per second}\right)^2}{64.4}$$

$$\text{Water horsepower} = \frac{\text{GPM} \times \text{head in ft.} \times \text{specific gravity}}{3960}$$

$$\text{Brake horsepower (pump)} = \frac{\text{GPM} \times \text{head in ft.} \times \text{specific gravity}}{3960 \times \text{pump efficiency}}$$

$$\text{Brake horsepower (motor)} = \frac{\text{Watts input} \times \text{motor efficiency}}{746}$$

$$\text{Efficiency (pump)} = \frac{\text{GPM} \times \text{head in ft.} \times \text{specific gravity}}{3960 \times \text{BHP}}$$

GENERAL PUMP FORMULAS (cont.)

$$\text{Motor KW} = \frac{\text{GPM x Head (ft.) x Specific Gravity x 0.746}}{3960 \times \text{Pump Efficiency} \times \text{Motor Efficiency}}$$

**Impeller Diameter
Change Formulas**

$$GPM_2 = \frac{D_2}{D_1} \times GPM_1$$

$$H_2 = \left(\frac{D_2}{D_1}\right)^2 \times H_1$$

$$BHP_2 = \left(\frac{D_2}{D_1}\right)^3 \times BHP_1$$

**Speed Change
Formulas**

$$GPM_2 = \frac{RPM_2}{RPM_1} \times GPM_1$$

$$H_2 = \left(\frac{RPM_2}{RPM_1}\right)^2 \times H_1$$

$$BHP_2 = \left(\frac{RPM_2}{RPM_1}\right)^3 \times BHP_1$$

Where: GPM = flow rate in gallons per minute
H = head in feet
BHP = brake horsepower
D = Impeller diameter
RPM = pump speed in revolutions per minute

SPECIFIC GRAVITY OF WATER

Temp. (F)	Specific Gravity	Temp. (F)	Specific Gravity
32°–100°	1.00	251°–300°	.92
101°–150°	.98	301°–350°	.90
151°–200°	.96	351°–400°	.87
201°–250°	.94	401°–450°	.83

DETERMINING PUMP MOTOR HORSEPOWER

Pump Flow (GPM)	Pump Pressure in PSI (80% efficiency)						
	25	50	75	100	250	500	1000
1	.02	.04	.05	.07	.18	.36	.72
2	.04	.07	.11	.14	.36	.72	1.45
3	.05	.10	.16	.21	.54	1.09	2.18
4	.08	.15	.22	.29	.72	1.45	2.91
5	.09	.18	.27	.36	.91	1.82	3.64
10	.18	.36	.54	.72	1.82	3.64	7.29
15	.27	.55	.82	1.09	2.73	5.46	10.93
20	.36	.73	1.09	1.45	3.64	7.29	14.58
25	.46	.91	1.37	1.82	4.55	9.11	18.23
30	.55	1.09	1.64	2.18	5.46	10.93	21.87
35	.64	1.28	1.91	2.55	6.38	12.76	25.52
40	.73	1.46	2.18	2.91	7.29	14.58	29.17
45	.82	1.64	2.46	3.28	8.20	16.40	32.81
50	.91	1.82	2.73	3.64	9.11	18.23	36.46
75	1.37	2.73	4.10	5.46	13.67	27.36	54.69
100	1.82	3.65	5.47	7.29	18.23	36.46	72.92

CENTRIFUGAL PUMP SPECIFICATIONS

Configuration of Pump	Maximum Flow Range (gpm)	Maximum TDH Range (ft.)	Maximum Pressure Range (psig)	Maximum Temperature Range (°F)
Horizontal, Overhung Impeller, Close Coupled, Single & Double Stage	30 to 15,000	10 to 5,500	20 to 5,000	-180 to 1,200
Horizontal, Overhung Impeller, Separately Coupled, Single & Double Stage	30 to 120,000	35 to 5,500	20 to 5,000	-400 to 1,500
Horizontal, Impeller Between Bearings, Single & Double Stage	5 to 180,000	20 to 2,600	75 to 3,000	-350 to 850
Horizontal, Impeller Between Bearings, Multi-Stage	20 to 140,000	260 to 15,000	140 to 7,000	-400 to 850
Vertical In-line, Overhung Impeller	20 to 140,000	15 to 6,900	100 to 5,000	-350 to 840
Vertical In-Line, Impeller Between Bearings	40 to 120,000	50 to 1,500	250 to 650	-65 to 350
Vertically Suspended, Single Stage, Separate Discharge	5 to 60,000	10 to 1,000	—	-150 to 1,200
Vertically Suspended, Multi-Stage, Discharge Through Column	500 to 600,000	50 to 7,200	—	-350 to 1,000
Vertically Suspended, Axial Flow	10,000 to 100,000	10 to 3,500	—	-50 to 1,000
Submersible, Single Stage	20 to 100,000	15 to 1,300	—	-100 to 2,000
Submersible, Multi-Single Stage Turbine	1,100 to 220,000	100 to 4,000	—	-50 to 700

ROTARY PUMP SPECIFICATIONS

Configuration of Pump	Maximum Flow Range (gpm)	Maximum TDH Range (ft.)	Maximum Pressure Range (psig)	Maximum Temperature Range (°F)
Vane, Flexible and Sliding	10 to 1,000	—	20 to 350	-100 to 600
Gear, External and Internal	5 to 120,000	—	90 to 10,000	-100 to 950
Piston	680 to 1,800	—	300 to 20,000	-150 to 800
Lobe	40 to 3,000	—	100 to 450	-40 to 500
Progressing Cavity	5 to 4,000	—	125 to 2,100	-20 to 1,300
Peristaltic	5 to 410	—	20 to 690	-60 to 300
Screw, Two and Three	35 to 15,000	—	580 to 6,900	-40 to 800

5-8

Configuration of Pump	Maximum Flow Range (gpm)	Maximum TDH Range (ft.)	Maximum Pressure Range (psig)	Maximum Temperature Range (°F)
RECIPROCATING PUMP SPECIFICATIONS				
Power, Plunger or Piston	5 to 5,000	—	20 to 100,000	-300 to 700
Power, Diaphragm	5 to 3,000	25 to 7,600	10 to 17,200	-300 to 850
Metering Reciprocating, Controlled Volume	5 to 18,200		5 to 60,000	—
SPECIAL PURPOSE PUMP SPECIFICATIONS				
Sanitary	5 to 2,000	—	5 to 14,000	-70 to 1,450
Non-electric, gas or vapor driven	5 to 5,000	20 to 23,000	15 to 10,000	-40 to 650
Cryogenic	5 to 21,400	25 to 2,500	600 to 17,000	-410 to 890
Hydrostatic Test	5 to 300	—	1,200 to 100,000	-40 to 600
Drum/carboy	5 to 220	15 to 290	—	-50 to 325

WELL PUMP CHARACTERISTICS

Pump Type	Speed	Suction Lift	Pressure Head	Delivery
Reciprocating Shallow well Low pressure Medium pressure	Slow 250 to 550 strokes per minute	22 to 25 ft.	40 to 43 psi Up to 100 psi	Pulsating Pulsating
High pressure			Up to 350 psi	Pulsating
Deep well	Slow 30 to 50 strokes per minute	Up to 875 ft. (Suction lift below cylinder 22 ft.)	40 psi (normal)	Pulsating
Rotary Pump Shallow well	400 to 1725 rpm	22 ft.	100 psi (normal)	Slightly pulsating
Ejector pump Shallow well and limited deep wells	Used with centrifugal turbine or shallow well reciprocating pump	120 ft. (maximum) 80 ft. (normal)	40 psi (normal) 70 psi (maximum)	Continuous, nonpulsating, high capacity, low-pressure head
Centrifugal Shallow well Single stage	High 1750 and 3600 rpm	15 ft. (maximum)	40 psi (normal) 70 psi (maximum)	Continuous, non- pulsating, high capacity, low-pressure head
Turbine Type Single impeller	High 1750 rpm	28 ft. (maximum at sea level)	40 psi (normal) 100 psi (maximum)	Continuous, non- pulsating, high capacity, low-pressure head

SUBMERSIBLE PUMP SELECTION GUIDELINES

Vertical Lift/Elevation

This is the overall vertical distance in feet between the well head and the level where the highest point of water use is located. For example, the vertical distance between the well head and the uppermost shower fixture.

Service Pressure

The service pressure is the range of pressure in the pressure tank during a typical pumping cycle. For example, if a pressure switch with a cut in/cut out pressure rating of 30 psi/50 psi is used, the range of pressure is 40 psi.

Pumping Level

The pumping level is the lowest water level reached in the well during a typical pumping cycle. To measure the pumping level in a well you must first determine the following:

a. Obtain the static (standing) water level in the well. This is the level of water in feet before pumping.

b. Measure the draw down of the well. This would be the distance in feet that the static water level in the well is lowered by pumping.

Thus, the static water level minus the draw down will equal the pumping level of the well.

Friction Loss

This is the loss of pressure measured in feet of head due to the resistance to flow inherent in the pipe and fittings.

Total Dynamic Head (TDH)

The TDH is the sum of the vertical lift, service pressure, pumping level and friction loss and is expressed in feet. TDH and the capacity required determines the pump size.

TROUBLESHOOTING – JET PUMPS

Pump Does Not Start or Run

Problem	Analysis	Correction
Bad start capacitor	Use ohmmeter to check resistance across capacitor. When contact is made, needle should jump. No movement indicates an open capacitor; no resistance indicates capacitor is shorted.	Replace capacitor.
Blown fuse	Check across each fuse with voltmeter. Voltage indicates blown fuse.	Replace fuse.
Loose connections, damaged wires or incorrect wiring	Check wiring diagram. See that all connections are tight and no short circuits exist.	Tighten connections. Replace damaged wires. Rewire incorrect circuits.
Low line voltage	Use voltmeter to check terminals nearest pump.	Check main switch on property. If voltage is low, contact power company. If not, larger wire may be required.
Bad motor	Make sure that switch is closed and motor is energized.	Repair or replace motor.

TROUBLESHOOTING – JET PUMPS *(cont.)*

Pump Does Not Start or Run

Problem	Analysis	Correction
Motor shorted	Fuse blowing when pump is started indicates motor is shorted or locked.	Repair or replace motor.
Defective pressure switch	Check switch setting. Examine switch contacts for dirt or excessive wear.	Adjust switch settings. Clean contacts if necessary. Replace switch.
Tubing to pressure switch clogged	Remove tubing and check for clog.	Clean or replace if clogged.
Impeller or seal binding	Turn off power. Try to turn impeller or motor with a long screwdriver.	If impeller does not turn, remove impeller housing and locate source of binding.

Overheating of Motor and Overload Trips

Problem	Analysis	Correction
Inadequate ventilation	Check air temperature around pump. If over 100°F, overload may be tripping due to external heat.	Provide adequate ventilation or move pump to another location.
Low pressure delivery	Continuous operation at low pressure results in overload on pump and may cause the overload protection to trip.	Install globe valve on discharge line and throttle to increase pressure.

TROUBLESHOOTING – JET PUMPS *(cont.)*

Overheating of Motor and Overload Trips

Problem	Analysis	Correction
Incorrect line voltage	Use voltmeter to check terminals nearest pump.	If voltage under recommended minimum, check wire size. If ok, contact power company.
Motor wired incorrectly	Check motor wiring diagram.	Rewire per wiring diagram.

Pump Starts and Stops Constantly

Problem	Analysis	Correction
Foot valve leak	Examine foot valve.	Repair or replace foot valve.
Suction side leak	Apply about 30 psi pressure to system. If system does not hold this pressure when compressor is off, the suction side is leaking.	Make sure above-ground connections are tight. Then repeat test. If necessary, pull piping and repair leak(s).
Discharge side leak	Shut all fixtures off. Check all components for leaks.	Repair leak(s).
Pressure tank leak	Apply soapy water to entire surface above water line. If bubbles appear, air is escaping from tank.	Repair leak(s). Replace tank.

TROUBLESHOOTING – JET PUMPS *(cont.)*

Pump Starts and Stops Constantly

Problem	Analysis	Correction
Defective pressure switch	Check switch setting. Examine switch contacts for dirt or excessive wear.	Adjust switch settings. Clean contacts. Replace switch.
Faulty air volume control	Will lead to a water-logged tank. Check that control is operating correctly. If not, remove and examine for clogging.	Clean or unclog control. Replace control.

Pump Does Not Shut Off

Problem	Analysis	Correction
Faulty pressure switch	Switch contacts may be welded together in closed position.	Replace switch.
Pressure switch is set incorrectly	Lower switch setting. If pump shuts off, bad setting is verified.	Adjust switch to proper setting.
Loss of prime	Check prime of piping and pump when water not delivered.	Re-prime.

TROUBLESHOOTING – JET PUMPS *(cont.)*

Pump Operates With Little or No Water Delivery

Problem	Analysis	Correction
Low well level	Compare well depth against the pump performance table. Be sure pump and ejector are sized properly.	Replace pump or replace ejector if not the correct size.
Low well capacity	Shut off pump and allow well to recover. Restart pump and determine whether delivery drops after continuous operation.	For a weak well, either lower the ejector (deep well), use a tall pipe or change from shallow well to deep well equipment
Piping undersized	If system delivery is low, lines may be undersized. Check friction loss.	Replace undersized piping. Install new larger capacity pump.
Clogged ejector	Remove ejector and inspect.	Clean and reinstall.
Pressure regulating valve stuck or incorrectly set (Deep well pumps)	Check valve setting. Check valve for dirt or defects.	Reset valve setting. Clean valve. Replace valve.
Air volume control leak	Disconnect control tubing at pump and plug hole. If capacity increases, leak exists in the tubing or control.	Tighten all fittings. Replace control if defective.

TROUBLESHOOTING – JET PUMPS *(cont.)*

Pump Operates With Little or No Water Delivery

Problem	Analysis	Correction
Suction side leak	Apply 30 psi pressure to system. If system will not hold pressure when compressor is off, the suction side is leaking.	Make sure above-ground connections are tight. Then repeat test. If necessary, pull piping and repair leak.
Incorrect pump and ejector combination	Check pump and ejector models against manufacturer performance tables.	Replace ejector if incorrect model.
Defective or clogged foot valve or strainer	Inspect foot valve. Partial clogging reduces delivery, complete clogging results in no delivery. Clogged foot valve or strainer causes pump to lose prime.	Clean both the foot valve and strainer. Repair or replace as required
Worn or defective parts. Plugged impeller	Disassemble parts and inspect impeller.	Replace worn or defective parts. Replace entire pump if necessary. Clean plugged impeller and if damaged, replace.
Loss of prime	Check prime of piping and pump when water not delivered.	Re-prime.

TROUBLESHOOTING – SUBMERSIBLE PUMPS

When Motor is Started, Breaker Trips or Fuses Blow

Problem	Analysis	Correction
Defective wiring	Check all motor and power wiring. Check that all connections are tight and no short circuits exist.	Rewire incorrect circuits. Tighten any loose connections. Replace worn or defective wires.
Incorrect components	Check all control components to insure they are correct type and size. Check service records.	Remove any incorrect components and replace with size and type recommended by the manufacturer.
Incorrect line voltage	Check line voltage at terminals.	If the voltage is incorrect, contact power company.
Defective starting capacitor	Use ohmmeter to determine resistance across capacitor. When contact is made, the needle should jump immediately, then creep up slowly. No movement indicates an open capacitor or bad relay points. No resistance indicates a shorted capacitor.	Replace defective capacitor.

TROUBLESHOOTING – SUBMERSIBLE PUMPS *(cont.)*

When Motor is Started,
Breaker Trips or Fuses Blow

Problem	Analysis	Correction
Defective pressure switch	Check voltage across switch. If less than line voltage, low voltage may be caused by bad contacts.	Clean contacts. Replace pressure switch.
Defective relay	Use ohmmeter to check relay coil. Resistance should be as shown in manufacturer specifications.	If coil resistance is incorrect or contacts defective, replace relay.
Shorted or open motor winding	Use ohmmeter to check resistance of motor winding. If resistance is too low, the motor winding may be shorted out. If resistance is too high or infinite, the motor winding is open.	If the motor winding is defective, the pump must be pulled and the motor repaired or replaced.
Grounded cable or winding	Ground one lead of ohmmeter to casing, touch other lead to each motor wire terminal. If needle moves significantly, there is a ground in either cable or motor winding.	Pull pump and inspect cable. Replace damaged cable. If cable is functioning properly, the motor winding is grounded.

TROUBLESHOOTING – SUBMERSIBLE PUMPS (*cont.*)

When Motor is Started, Breaker Trips or Fuses Blow

Problem	Analysis	Correction
Sandlocked pump	Reverse pump flow by interchanging main and start winding leads at control box.	Pull pump, disassemble and clean. Be sure sand has settled in well. Rewire and reinstall pump.
Crooked well	Pump may have become misaligned resulting in a locked rotor.	Pull pump and straighten well. Realign pump and reinstall.

Pump Operates With Little or No Water Delivery

Problem	Analysis	Correction
Discharge line check valve installed backward	Examine check valve on discharge line for proper flow direction.	Reverse valve if required.
Drop pipe leak	Raise drop pipe and inspect all sections.	Replace defective section(s).
Pump check valve jammed	Pull the pump. Examine drop pipe connection to pump outlet. If threaded section of pipe is screwed in too far, it may have jammed the check valve in the closed position.	Unscrew drop pipe and cut off a small section of the threads. After assembly, check-valve flapper should operate freely.

TROUBLESHOOTING – SUBMERSIBLE PUMPS (*cont.*)

Pump Operates With Little or No Water Delivery

Problem	Analysis	Correction
Intake screen blocked	Mud or sand may be blocking screen. Pull pump and check screen.	Clean screen. Be sure to reinstall pump three to four feet above bottom of well.
Motor shaft loose	Coupling between motor and shaft may have become loose. Pull pump and examine coupling and shaft for wear.	Tighten all loose connections. Replace components as necessary.
Pump parts worn	Abrasives in water may result in excessive wear. Before pulling pump, reduce setting on pressure switch to determine if pump shuts off. If it does, examine pump for worn parts.	Pull pump and replace worn components.
Air locked pump	Stop and start pump several times. Wait between cycles. If pump resumes normal delivery, air lock was the problem.	If cycling off and on fails to correct the problem, partially restrict the setting on the valve or cock.
Loss of Prime	Check prime of piping and pump when water not delivered.	Re-prime.

TROUBLESHOOTING – SUBMERSIBLE PUMPS (cont.)

Pump Operates With Little or No Water Delivery

Problem	Analysis	Correction
Well water level too low	Well production may be too low for capacity of pump. Partially restrict pump output, and allow well to recover. Restart pump.	If partial restriction corrects the problem, lock in the valve or cock at restricted setting. Otherwise, lower pump in recovered well.

Pump Starts Constantly

Problem	Analysis	Correction
Discharge line check valve leaking	Remove valve and examine for defects.	Replace valve.
Pressure tank leak (above water level)	Apply soap solution to entire tank surface. Bubbles indicate air is escaping tank.	Repair or replace tank.
Plumbing system leak	Examine service line and distribution branches for leaks.	Repair leaks.
Snifter valve clogged	Remove and inspect snifter valve.	Clean or replace valve.
Air volume control clogged	Remove and inspect air volume control.	Clean or replace control.
Pressure switch defective or out of adjustment	Check pressure setting. Examine switch for defects.	Reduce pressure setting. Replace switch.

TROUBLESHOOTING—SUBMERSIBLE PUMPS (*cont.*)

Pump Does Not Shut Off

Problem	Analysis	Correction
Drop line leak	Raise drop pipe and check for leaks.	Replace damaged section(s) of pipe.
Well water level too low	Well production may be too low for capacity of pump. Partially restrict pump output and allow well to recover. Restart pump.	If partial restriction corrects problem, lock in the valve or cock at restricted setting. Otherwise, lower pump in recovered well.
Defective pressure switch	Check points for dirt. Examine other parts of switch for defects or excessive wear.	Clean points. Replace switch.
Pump parts worn	Abrasives in water may result in excessive wear on pump parts such as the casing and the impeller. Reduce setting on pressure switch to determine if pump shuts off. If it does, examine pump for worn parts.	Pull pump and replace worn components.

TROUBLESHOOTING – SUBMERSIBLE PUMPS (*cont.*)

Motor Does Not Start, Yet Fuses Are Not Blowing

Problem	Analysis	Correction
No overload protection	Check fuses or circuit breaker to be sure they are functioning.	Replace blown fuses. Reset tripped breaker. Replace if necessary.
No power	Use voltmeter to check power supply across incoming power lines.	If no power is going to the control box, contact power company.
Defective control box	Examine wiring in box to be sure contacts are secure. Check voltage at motor terminals. If no voltage is indicated, wiring is defective.	Tighten loose contacts. Replace defective wiring.
Defective pressure switch	Check voltage across switch while switch is in closed position. If voltage drop is equal to the line voltage, the switch is not making contact.	Clean points. Replace switch.

TROUBLESHOOTING – SUBMERSIBLE PUMPS (*cont.*)

Blowing Fuses While Motor is Running

Problem	Analysis	Correction
Sandlocked pump	If fuses blow while pump is operating, sand or dirt may have jammed the impeller, causing a locked rotor.	Pull pump, disassemble and clean. Before replacing, make sure the sand in the well has settled.
Defective motor winding or cable	Use an ohmmeter to check resistance of the motor winding. Resistance should match the specifications indicated in the manufacturer's literature. If too low, the motor winding may be shorted. If there is no needle movement, there is high or infinite resistance indicating an open circuit in the motor winding. Ground one lead of the ohmmeter onto the drop pipe or shell casing, then apply the other lead to each motor terminal. If the needle moves substantially, there is a ground in either the motor winding or the cable.	If neither the cable nor the winding is shorted, grounded or open (defective), then the pump must be pulled and serviced.

TROUBLESHOOTING – SUBMERSIBLE PUMPS (cont.)

Blowing Fuses While Motor is Running

Problem	Analysis	Correction
Defective control box components	Use ohmmeter to determine resistance across the running capacitor. When contact is made, ohmmeter needle should jump immediately, then creep up slowly. No movement indicates an open capacitor. No resistance means a shorted capacitor. Use ohmmeter to check relay coil. Resistance should match the specifications as indicated in the manufacturer's literature.	Replace defective control box components.
Overheated overload protection box	If box is hot to the touch, this may be the problem.	Ventilate, shade or move box to eliminate source of heat.
Incorrect voltage	Use a voltmeter to check line voltage terminals. Be sure that the voltage is within the range specified by the manufacturer.	If voltage incorrect, contact power company.

CHAPTER 6
Excavation

SOIL TYPES

Type A soils are cohesive soils with an unconfined, compressive strength of 1.5 tons per square foot (tsf) (144kPa) or greater. Examples of Type A soils include clay, silty clay, sandy clay, clay loam and, in some cases, silty clay loam and sandy clay loam. Cemented soils such as caliche and hardpan are also considered Type A. However, no soil is Type A if:

- The soil is fissured.
- The soil is subject to vibration from heavy traffic, pile driving, or similar effects.
- The soil has been previously disturbed.
- The soil is part of a sloped, layered system where the layers dip into the excavation on a slope of 4 horizontal to 1 vertical (4H:1V) or greater.
- The material is subject to other factors that would require it to be classified as a less stable material.

Type B soils are cohesive soils with an unconfined, compressive strength greater than 0.5 tsf (48 kPa) but less than 1.5 tsf (144 kPa). Examples of Type B soils include angular gravel (similar to crushed rock), silt, silt loam, sandy loam and, in some cases, silty clay loam and sandy clay loam; previously disturbed soils

---

except those which would otherwise be classified as Type C soils; soils that meet the unconfined, compressive strength or cementation requirements for Type A, but are fissured or subject to vibration; dry rock that is not stable, and material that is part of a sloped, layered system where the layers dip into the excavation on a slope less steep than 4 horizontal to 1 vertical (4H:1V), but only if the material would otherwise be classified as Type B.

Type C soils are cohesive soils with an unconfined, compressive strength of 0.5 tsf (48 kPa) or less. Examples of Type C soils include granular soils such as gravel, sand and loamy sand; submerged soil or soil from which water is freely seeping; submerged rock that is not stable, and material in a sloped, layered system where the layers dip into the excavation or a slope of 4 horizontal to 1 vertical (4H:1V) or steeper.

Unconfined compressive strength means the load per unit area at which a soil will fail in compression. It can be determined by laboratory testing or estimated in the field using a pocket penetrometer or by thumb penetration tests and other methods.

Wet soil means soil that contains significantly more moisture than moist soil, but in such a range of values that cohesive material will slump or begin to flow when vibrated. Granular material that would exhibit cohesive properties when moist will lose those cohesive properties when wet.

TERMS USED IN GRADING

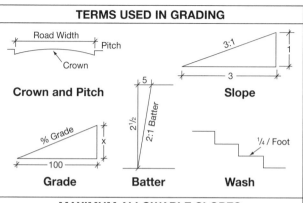

Crown and Pitch

Slope

Grade

Batter

Wash

MAXIMUM ALLOWABLE SLOPES

Soil or Rock Type	Maximum Allowable Slopes (H:V)[1] for Excavations Less Than 20 Feet Deep[3]
Stable Rock	Vertical (90 degrees)
Type A[2]	¾:1 (53 degrees)
Type B	1:1 (45 degrees)
Type C	1½:1 (34 degrees)

[1]Numbers shown in parentheses next to maximum allowable slopes are angles expressed in degrees from the horizontal. Angles have been rounded off.

[2]A short-term maximum allowable slope of ½H:1V (63 degrees) is allowed in excavations in Type A soil that are 12 feet (3.67 m) or less in depth. Short-term maximum allowable slopes for excavations greater than 12 feet (3.67 m) in depth shall be ¾H:1V (53 degrees).

[3]Sloping or benching for excavations greater than 20 feet deep should be designed by a registered professional engineer.

EXCAVATING TOWARD AN OPEN DITCH

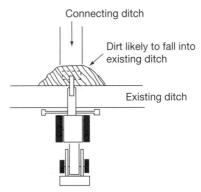

Hazard From Falling Dirt

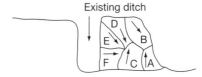

Excavate in This Sequence to Avoid Falling Dirt

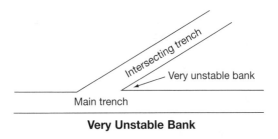

Very Unstable Bank

SEQUENCE OF EXCAVATION IN HIGH GROUNDWATER

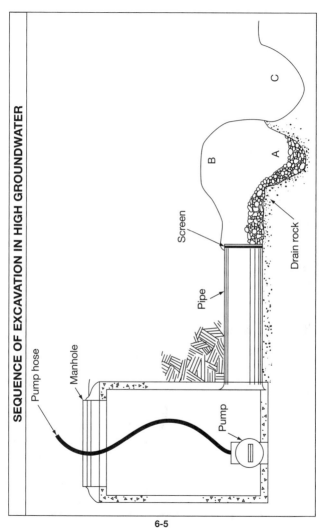

Pump hose

Manhole

Screen

B

C

A

Pipe

Drain rock

Pump

6-5

ANGLE OF REPOSE FOR COMMON SOIL TYPES

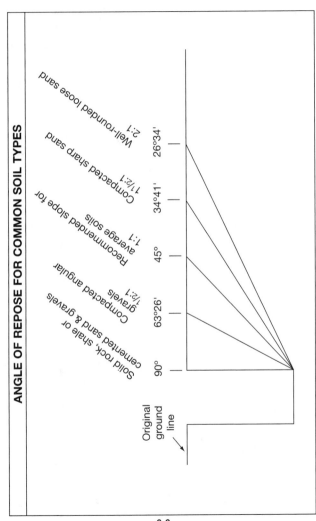

Solid rock, shale or cemented sand & gravels — 90°

Compacted angular gravels — 1/2:1 — 63°26'

Recommended slope for average soils — 1:1 — 45°

Compacted sharp sand — 1 1/2:1 — 34°41'

Well-rounded loose sand — 2:1 — 26°34'

Original ground line

6-6

CONE OF DEPRESSION

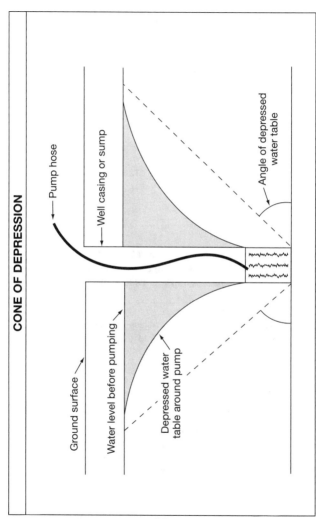

Pump hose

Well casing or sump

Angle of depressed water table

Ground surface

Water level before pumping

Depressed water table around pump

EXCAVATIONS MADE IN TYPE A SOIL

All simple slope excavations 20' or less in depth shall have a maximum allowable slope of ¾:1.

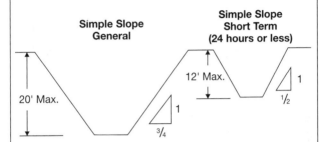

Simple Slope General

Simple Slope Short Term (24 hours or less)

20' Max.

12' Max.

¾ / 1

½ / 1

Exception: Simple slope excavations which are open 24 hours or less (short term) and which are 12' or less in depth shall have a maximum allowable slope of ½:1.

All benched excavations 20' or less in depth shall have a maximum allowable slope of ¾ to 1 and maximum bench dimensions as follows:

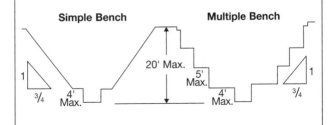

Simple Bench

Multiple Bench

1 / ¾

4' Max.

20' Max.

5' Max.

4' Max.

¾ / 1

EXCAVATIONS MADE IN TYPE B SOIL

All simple slope excavations 20' or less in depth shall have a maximum allowable slope of 1:1.

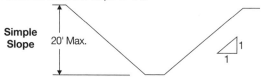

Simple Slope

20' Max.

All benched excavations 20' or less in depth shall have a maximum allowable slope of 1:1 and maximum bench dimensions as follows:

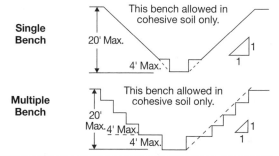

Single Bench

This bench allowed in cohesive soil only.

20' Max.

4' Max.

Multiple Bench

This bench allowed in cohesive soil only.

20' Max.

4' Max.

4' Max.

All excavations 20' or less in depth which have vertically sided lower portions shall be shielded or supported to a height at least 18" above the top of the vertical side. All such excavations shall have a maximum allowable slope of 1:1.

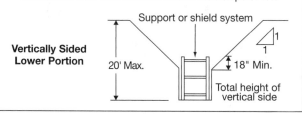

Vertically Sided Lower Portion

Support or shield system

20' Max.

18" Min.

Total height of vertical side

EXCAVATIONS MADE IN TYPE C SOIL

All simple slope excavations 20' or less in depth shall have a maximum allowable slope of 1½:1.

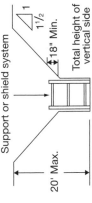

Simple Slope

20' Max.

1
1½

All excavations 20' or less in depth which have vertically sided lower portions shall be shielded or supported to a height at least 18" above the top of the vertical side. All such excavations shall have a maximum allowable slope of 1½:1.

Vertically Sided Lower Portion

Support or shield system

20' Max.

1
1½

18" Min.

Total height of vertical side

All excavations 8' or less in depth which have unsupported vertically sided lower portions shall have a maximum vertical side of 3½'.

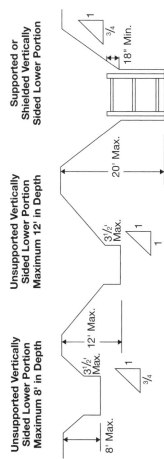

Unsupported Vertically Sided Lower Portion Maximum 8' in Depth

3½' Max.

8' Max.

1 / ¾

Unsupported Vertically Sided Lower Portion Maximum 12' in Depth

3½' Max.

1 / 1

12' Max.

20' Max.

Supported or Shielded Vertically Sided Lower Portion

18" Min.

1 / ¾

All excavations more than 8' but not more than 12' in depth with unsupported vertically sided lower portions shall have a maximum allowable slope of 1:1 and a maximum vertical side of 3½'.

All excavations 20' or less in depth which have vertically sided lower portions that are supported or shielded shall have a maximum allowable slope of ¾:1. The support or shield system must extend at least 18" above the top of the vertical side.

6-11

EXCAVATIONS MADE IN LAYERED SOILS

All excavations 20' or less in depth made in layered soils have a maximum allowable slope for each layer.

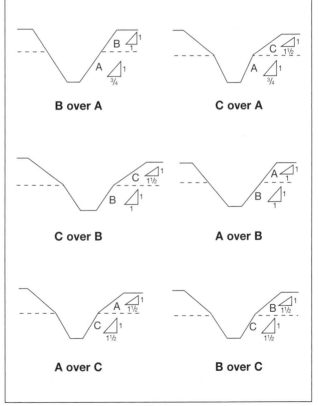

B over A

C over A

C over B

A over B

A over C

B over C

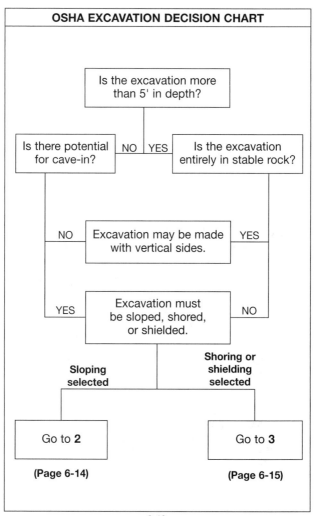

OSHA EXCAVATION DECISION CHART

Is the excavation more than 5' in depth?

Is there potential for cave-in? — NO | YES — Is the excavation entirely in stable rock?

NO — Excavation may be made with vertical sides. — YES

YES — Excavation must be sloped, shored, or shielded. — NO

Sloping selected

Shoring or shielding selected

Go to **2**

(Page 6-14)

Go to **3**

(Page 6-15)

6-13

```
                    ┌─────────────────────────┐
                    │            2            │
                    │  Sloping selected as the │
                    │   method of protection  │
                    └─────────────────────────┘
                                 │
                    ┌─────────────────────────┐
                    │ Will soil classification be │
                    │ made in accordance with │
                    │      §1926.652 (b)?     │
                    └─────────────────────────┘
```

YES   NO

Excavation must comply   Excavations must
with one of the following   comply with
three options: §1926.652 (b)(1)
                                which requires
Option 1:   a slope of
§1926.652 (b)(2) which   1½H:1V (34°).
requires Appendices A and
B to be followed.

Option 2:
§1926.652 (b)(3) which
requires other tabulated
data to be followed.

Option 3:
§1926.652 (b)(4) which
requires the excavation to
be designed by a registered
professional engineer.

3
Shoring or shielding
selected as the method
of protection

Soil classification is required when
shoring or shielding is used.
The excavation must comply with
one of the following four options:

Option 1:
§1926.652 (c)(1) which requires
Appendices A and C to be followed
(e.g., timber shoring).

Option 2:
§1926.652 (c)(2) which requires
manufacturers data to be followed
(e.g., hydraulic shoring, trench jacks,
air shores, shields).

Option 3:
§1926.652 (c)(3) which requires tabulated
data to be followed (e.g., any system
as per the tabulated data).

Option 4:
§1926.652 (c)(4) which requires
the excavation to be designed by a
registered professional engineer
(e.g., any designed system).

TIMBER TRENCH SHORING
MINIMUM TIMBER REQUIREMENTS — SOIL TYPE A

P(a) = 25 x H + 72 psf (2 ft. surcharge)

Depth of Trench (ft.)	Size (actual) and Spacing of Members						
	Cross Braces						
	Horiz. Spacing (ft.)	Width of Trench (ft.)					Vertical Spacing (ft.)
		Up to 4	Up to 6	Up to 9	Up to 12	Up to 15	
5 to 10	Up to 6	4 x 4	4 x 4	4 x 6	6 x 6	6 x 6	4
	Up to 8	4 x 4	4 x 4	4 x 6	6 x 6	6 x 6	4
	Up to 10	4 x 6	4 x 6	4 x 6	6 x 6	6 x 6	4
	Up to 12	4 x 6	4 x 6	6 x 6	6 x 6	6 x 6	4
10 to 15	Up to 6	4 x 4	4 x 4	4 x 6	6 x 6	6 x 6	4
	Up to 8	4 x 6	4 x 6	6 x 6	6 x 6	6 x 6	4
	Up to 10	6 x 6	6 x 6	6 x 6	6 x 8	6 x 8	4
	Up to 12	6 x 6	6 x 6	6 x 6	6 x 8	6 x 8	4
15 to 20	Up to 6	6 x 6	6 x 6	6 x 6	6 x 8	6 x 8	4
	Up to 8	6 x 6	6 x 6	6 x 6	6 x 8	6 x 8	4
	Up to 10	8 x 8	8 x 8	8 x 8	8 x 8	8 x 10	4
	Up to 12	8 x 8	8 x 8	8 x 8	8 x 8	8 x 10	4

TIMBER TRENCH SHORING *(cont.)*
MINIMUM TIMBER REQUIREMENTS — SOIL TYPE A

P(a) = 25 x H + 72 psf (2 ft. surcharge)

Depth of Trench (ft.)	Size (actual) and Spacing of Members						
	Wales		Uprights				
			Maximum Allowable Horizontal Spacing (ft.)				
	Size (in.)	Vertical Spacing (ft.)	Close	4	5	6	8
5 to 10	Not req'd.	–	–	–	–	2 x 6	–
	Not req'd.	–	–	–	–	–	2 x 8
	8 x 8	4	–	–	2 x 6	–	–
	8 x 8	4	–	–	–	2 x 6	–
10 to 15	Not req'd.	–	–	–	–	3 x 8	–
	8 x 8	4	–	2 x 6	–	–	–
	8 x 10	4	–	–	2 x 6	–	–
	10 x 10	4	–	–	–	3 x 8	–
15 to 20	6 x 8	4	3 x 6	–	–	–	–
	8 x 8	4	3 x 6	–	–	–	–
	8 x 10	4	3 x 6	–	–	–	–
	10 x 10	4	3 x 6	–	–	–	–

Mixed oak or equivalent with a bending strength not less than 850 psi. Manufactured members of equivalent strength may be substituted for wood.

TIMBER TRENCH SHORING
MINIMUM TIMBER REQUIREMENTS — SOIL TYPE B

$P(a) = 45 \times H + 72$ psf (2 ft. surcharge)

Depth of Trench (ft.)	Size (actual) and Spacing of Members						
	Cross Braces						
	Horiz. Spacing (ft.)	Width of Trench (ft.)					Vertical Spacing (ft.)
		Up to 4	Up to 6	Up to 9	Up to 12	Up to 15	
5 to 10	Up to 6	4 x 6	4 x 6	6 x 6	6 x 6	6 x 6	5
	Up to 8	6 x 6	6 x 6	6 x 6	6 x 8	6 x 8	5
	Up to 10	6 x 6	6 x 6	6 x 6	6 x 8	6 x 8	5
10 to 15	Up to 6	6 x 6	6 x 6	6 x 6	6 x 8	6 x 8	5
	Up to 8	6 x 8	6 x 8	6 x 8	8 x 8	8 x 8	5
	Up to 10	8 x 8	8 x 8	8 x 8	8 x 8	8 x 10	5
15 to 20	Up to 6	6 x 8	6 x 8	6 x 8	8 x 8	8 x 8	5
	Up to 8	8 x 8	8 x 8	8 X 8	8 x 8	8 x 10	5
	Up to 10	8 x 10	8 x 10	8 X 10	8 x 10	10 x 10	5

TIMBER TRENCH SHORING *(cont.)*
MINIMUM TIMBER REQUIREMENTS — SOIL TYPE B

$P(a) = 45 \times H + 72$ psf (2 ft. surcharge)

Depth of Trench (ft.)	Size (actual) and Spacing of Members				
	Wales		Uprights		
			Maximum Allowable Horizontal Spacing (ft.)		
	Size (in.)	Vertical Spacing (ft.)	Close	2	3
5 to 10	6 x 8	5	–	–	2 x 6
	8 x 10	5	–	–	2 x 6
	10 x 10	5	–	–	2 x 6
10 to 15	8 x 8	5	–	2 x 6	–
	10 x 10	5	–	2 x 6	–
	10 x 12	5	–	2 x 6	–
15 to 20	8 x 10	5	3 x 6	–	–
	10 x 12	5	3 x 6	–	–
	12 x 12	5	3 x 6	–	–

Mixed oak or equivalent with a bending strength not less than 850 psi. Manufactured members of equivalent strength may be substituted for wood.

TIMBER TRENCH SHORING
MINIMUM TIMBER REQUIREMENTS — SOIL TYPE C

P(a) = 80 x H + 72 psf (2 ft. surcharge)

Depth of Trench (ft.)	Horiz. Spacing (ft.)	Up to 4	Up to 6	Up to 9	Up to 12	Up to 15	Vertical Spacing (ft.)
		Size (actual) and Spacing of Members					
		Cross Braces					
		Width of Trench (ft.)					
5 to 10	Up to 6	6 x 8	6 x 8	6 x 8	8 x 8	8 x 8	5
	Up to 8	8 x 8	8 x 8	8 x 8	8 x 8	8 x 10	5
	Up to 10	8 x 10	8 x 10	8 x 10	8 x 10	10 x 10	5
10 to 15	Up to 6	8 x 8	8 x 8	8 x 8	8 x 8	8 x 10	5
	Up to 8	8 x 10	8 x 10	8 x 10	8 x 10	10 x 10	5
15 to 20	Up to 6	8 x 10	8 x 10	8 x 10	8 x 10	10 x 10	5

TIMBER TRENCH SHORING *(cont.)*
MINIMUM TIMBER REQUIREMENTS — SOIL TYPE C

P(a) = 80 x H + 72 psf (2 ft. surcharge)

Depth of Trench (ft.)	Size (actual) and Spacing of Members		
	Wales		Uprights
	Size (in.)	Vertical Spacing (ft.)	Maximum Allowable Horizontal Spacing (ft.)
			Close
5 to 10	8 x 10	5	2 x 6
	10 x 12	5	2 x 6
	12 x 12	5	2 x 6
10 to 15	10 x 12	5	2 x 6
	12 x 12	5	2 x 6
15 to 20	12 x 12	5	3 x 6

Mixed oak or equivalent with a bending strength not less than 850 psi. Manufactured members of equivalent strength may be substituted for wood.

TIMBER TRENCH SHORING
MINIMUM TIMBER REQUIREMENTS — SOIL TYPE A

$P(a) = 25 \times H + 72$ psf (2 ft. surcharge)

Depth of Trench (ft.)	Size (S4S) and Spacing of Members						
	Cross Braces						
	Horiz. Spacing (ft.)	Width of Trench (ft.)					Vertical Spacing (ft.)
		Up to 4	Up to 6	Up to 9	Up to 12	Up to 15	
5 to 10	Up to 6	4 x 4	4 x 4	4 x 4	4 x 4	4 x 6	4
	Up to 8	4 x 4	4 x 4	4 x 4	4 x 6	4 x 6	4
	Up to 10	4 x 6	4 x 6	4 x 6	6 x 6	6 x 6	4
	Up to 12	4 x 6	4 x 6	4 x 6	6 x 6	6 x 6	4
10 to 15	Up to 6	4 x 4	4 x 4	4 x 4	6 x 6	6 x 6	4
	Up to 8	4 x 6	4 x 6	4 x 6	6 x 6	6 x 6	4
	Up to 10	6 x 6	6 x 6	6 x 6	6 x 6	6 x 6	4
	Up to 12	6 x 6	6 x 6	6 x 6	6 x 6	6 x 6	4
15 to 20	Up to 6	6 x 6	6 x 6	6 x 6	6 x 6	6 x 6	4
	Up to 8	6 x 6	6 x 6	6 x 6	6 x 6	6 x 6	4
	Up to 10	6 x 6	6 x 6	6 x 6	6 x 6	6 x 8	4
	Up to 12	6 x 6	6 x 6	6 x 6	6 x 8	6 x 8	4

TIMBER TRENCH SHORING (cont.)
MINIMUM TIMBER REQUIREMENTS — SOIL TYPE A

P(a) = 25 x H + 72 psf (2 ft. surcharge)

Depth of Trench (ft.)	Size (S4S) and Spacing of Members						
	Wales		Uprights				
	Size (in.)	Vertical Spacing (ft.)	Maximum Allowable Horizontal Spacing (ft.)				
			Close	4	5	6	8
5 to 10	Not req'd.	Not req'd.	–	–	–	4 x 6	–
	Not req'd.	Not req'd.	–	–	–	–	4 x 8
	8 x 8	4	–	–	4 x 6	–	–
	8 x 8	4	–	–	–	4 x 6	–
10 to 15	Not req'd.	Not req'd.	–	–	–	4 x 10	–
	6 x 8	4	–	4 x 6	–	–	–
	8 x 8	4	–	–	4 x 8	–	–
	8 x 10	4	–	4 x 6	–	4 x 10	–
15 to 20	6 x 8	4	3 x 6	–	–	–	–
	8 x 8	4	3 x 6	4 x 12	–	–	–
	8 x 10	4	3 x 6	–	–	–	–
	8 x 12	4	3 x 6	4 x 12	–	–	–

Douglas fir or equivalent with a bending strength not less than 1500 psi. Manufactured members of equivalent strength may be substituted for wood.

TIMBER TRENCH SHORING
MINIMUM TIMBER REQUIREMENTS — SOIL TYPE B

$P(a) = 45 \times H + 72$ psf (2 ft. surcharge)

Depth of Trench (ft.)	Size (S4S) and Spacing of Members						
	Cross Braces						
	Horiz. Spacing (ft.)	Width of Trench (ft.)					Vertical Spacing (ft.)
		Up to 4	Up to 6	Up to 9	Up to 12	Up to 15	
5 to 10	Up to 6	4 x 6	4 x 6	4 x 6	6 x 6	6 x 6	5
	Up to 8	4 x 6	4 x 6	6 x 6	6 x 6	6 x 6	5
	Up to 10	4 x 6	4 x 6	6 x 6	6 x 6	6 x 8	5
10 to 15	Up to 6	6 x 6	6 x 6	6 x 6	6 x 8	6 x 8	5
	Up to 8	6 x 8	6 x 8	6 x 8	8 x 8	8 x 8	5
	Up to 10	6 x 8	6 x 8	8 x 8	8 x 8	8 x 8	5
15 to 20	Up to 6	6 x 8	6 x 8	6 x 8	6 x 8	8 x 8	5
	Up to 8	6 x 8	6 x 8	6 x 8	8 x 8	8 x 8	5
	Up to 10	8 x 8	8 x 8	8 x 8	8 x 8	8 x 8	5

TIMBER TRENCH SHORING *(cont.)*
MINIMUM TIMBER REQUIREMENTS — SOIL TYPE B

P(a) = 45 x H + 72 psf (2 ft. surcharge)

Depth of Trench (ft.)	Size (S4S) and Spacing of Members				
	Wales		Uprights		
	Size (in.)	Vertical Spacing (ft.)	Maximum Allowable Horizontal Spacing (ft.)		
			Close	2	3
5 to 10	6 x 8	5	–	–	3 x 12 4 x 8
	8 x 8	5	–	3 x 8	
	8 x 10	5	–	–	4 x 8
10 to 15	8 x 8	5	3 x 6	4 x 10	–
	10 x 10	5	3 x 6	4 x 10	–
	10 x 12	5	3 x 6	4 x 10	–
15 to 20	8 x 10	5	4 x 6	–	–
	10 x 12	5	4 x 6	–	–
	12 x 12	5	4 x 6	–	–

Douglas fir or equivalent with a bending strength not less than 1500 psi. Manufactured members of equivalent strength may be substituted for wood.

TIMBER TRENCH SHORING
MINIMUM TIMBER REQUIREMENTS — SOIL TYPE C

P(a) = 80 x H + 72 psf (2 ft. surcharge)

Depth of Trench (ft.)	Horiz. Spacing (ft.)	Size (S4S) and Spacing of Members					Vertical Spacing (ft.)
		Cross Braces					
		Width of Trench (ft.)					
		Up to 4	Up to 6	Up to 9	Up to 12	Up to 15	
5 to 10	Up to 6	6 x 6	6 x 6	6 x 6	6 x 6	8 x 8	5
	Up to 8	6 x 6	6 x 6	6 x 6	8 x 8	8 x 8	5
	Up to 10	6 x 6	6 x 6	8 x 8	8 x 8	8 x 8	5
10 to 15	Up to 6	6 x 8	6 x 8	6 x 8	8 x 8	8 x 8	5
	Up to 8	8 x 8	8 x 8	8 x 8	8 x 8	8 x 8	5
15 to 20	Up to 6	8 x 8	8 x 8	8 x 8	8 x 10	8 x 10	5

TIMBER TRENCH SHORING (cont.)
MINIMUM TIMBER REQUIREMENTS — SOIL TYPE C

P(a) = 80 x H + 72 psf (2 ft. surcharge)

Depth of Trench (ft.)	Size (S4S) and Spacing of Members		Uprights
	Wales		
	Size (in.)	Vertical Spacing (ft.)	Maximum Allowable Horizontal Spacing (ft.)
			Close
5 to 10	8 x 8	5	3 x 6
	10 x 10	5	3 x 6
	10 x 12	5	3 x 6
10 to 15	10 x 10	5	4 x 6
	12 x 12	5	4 x 6
15 to 20	10 x 12	5	4 x 6

LOAD-CARRYING CAPACITIES OF SOILS

Type of soil	Tons per sq. ft.
Soft clay	1
Firm clay or fine sand	2
Compact fine or loose coarse sand	3
Loose gravel or compact coarse sand	4
Compact sand-gravel mixture	6

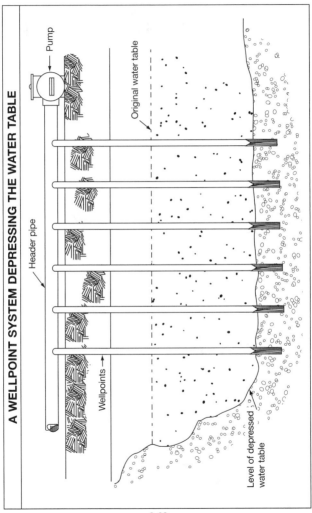

A WELLPOINT SYSTEM DEPRESSING THE WATER TABLE

Pump

Header pipe

Wellpoints

Original water table

Level of depressed water table

6-28

CHAPTER 7
General Construction and Safety

DRAWING SCALES			
Plan Use	**Ratio**	**Metric Length**	**English Equivalent (approx.)**
Details:	1:1	1000 mm = 1 m	12" = 1' 0" (full scale)
	1:5	200 mm = 1 m	3" = 1' 0"
	1:10	100 mm = 1 m	$1\frac{1}{2}$" = 1' 0"
	1:20	50 mm = 1 m	$\frac{1}{2}$" = 1' 0"
Floor plans:	1:40	25 mm = 1 m	$\frac{3}{8}$" = 1' 0"
	1:50	20 mm = 1 m	$\frac{1}{4}$" = 1' 0"
Plot plans:	1:80	13.3 mm = 1 m	$\frac{3}{16}$" = 1' 0"
	1:100	12.5 mm = 1 m	$\frac{1}{8}$" = 1' 0"
	1:200	5 mm = 1 m	1" = 20' 0"
Plat plans:	1:500	2 mm = 1 m	1" = 50' 0"
City maps (and larger):	1:1250	0.8 mm = 1 m	1" = 125' 0"
	1:2500	0.4 mm = 1 m	1" = 250' 0"

WIRE ROPE CHARACTERISTICS
FOR 6 STRAND BY 19 WIRE TYPE

WR Diameter (in.)	Weight (lbs./foot)	Breaking Point (lbs.)	Safe Load (lbs.)
$\frac{1}{4}$	0.10	4800	675
$\frac{5}{16}$	0.16	7400	1000
$\frac{3}{8}$	0.23	10600	1500
$\frac{7}{16}$	0.31	14400	2000
$\frac{1}{2}$	0.40	18700	2400
$\frac{9}{16}$	0.51	23600	3300
$\frac{5}{8}$	0.63	29000	4000
$\frac{3}{4}$	0.90	41400	6000
$\frac{7}{8}$	1.23	56000	8000
1	1.60	72800	10000
$1\frac{1}{8}$	2.03	91400	13000
$1\frac{1}{4}$	2.50	112400	16000
$1\frac{3}{8}$	3.03	135000	19000
$1\frac{1}{2}$	3.60	160000	22000
$1\frac{3}{4}$	4.90	216000	30500
2	6.40	278000	40000
$2\frac{1}{2}$	10.00	424000	60000

**The above values are for vertical pulls
at average ambient temperatures.**

CABLE CLAMPS PER WIRE ROPE SIZE

Wire Rope Diameter (in.)	Number of Clamps Required	Clip Spacing (in.)	Rope Turn-back (in.)
1/8	2	3	3 1/4
3/16	2	3	3 3/4
1/4	2	3 1/4	4 3/4
5/16	2	3 1/4	5 1/4
3/8	2	4	6 1/2
7/16	2	4 1/2	4
1/2	3	5	11 1/2
9/16	3	5 1/2	12
5/8	3	5 3/4	12
3/4	4	6 3/4	18
7/8	4	8	19
1	5	8 3/4	26
1 1/8	6	9 3/4	34
1 1/4	6	10 3/4	37
1 7/16	7	11 1/2	44
1 1/2	7	12 1/2	48
1 5/8	7	13 1/4	51
1 3/4	7	14 1/2	53
2	8	16 1/2	71
2 1/4	8	16 1/2	73
2 1/2	9	17 3/4	84
2 3/4	10	18	100
3	10	18	106

WIRE ROPE SLINGS – STRAIGHT LEG

Wire Rope Diameter (in.)	Straight 1 Leg	Choker 1 Leg	60° Choker 2 Leg	45° Choker 2 Leg	30° Choker 2 Leg
			CAPACITY IN TONS		
1/4	1/2	1/3	2/3	1/2	1/3
3/8	1	3/4	1 1/4	1	3/4
1/2	2	1 1/2	2 1/2	2	1 1/2
5/8	3	2	4	3	2
3/4	4	3	5	4	3
1	7	5	8	7	5
1 1/4	10	7	12	9	7
1 1/2	13	9	16	13	9
2	21	15	27	22	15
2 1/2	28	22	38	31	22
3	36	28	49	40	28
3 1/2	40	34	59	48	34

WIRE ROPE SLINGS – BASKET

Wire Rope Diameter (in.)	60° Basket 2 Leg	45° Basket 2 Leg	30° Basket 2 Leg	60° Basket 4 Leg	45° Basket 4 Leg	30° Basket 4 Leg
			CAPACITY IN TONS			
1/4	2/3	1/2	1/3	1	1	3/4
3/8	1 1/2	1	3/4	3	2	1 1/2
1/2	2 1/2	2	1 1/2	5	4	3
5/8	4	3	2	7	6	4
3/4	5	4	3	11	9	6
1	9	7	5	18	15	10
1 1/4	13	11	7	26	21	15
1 1/2	17	14	10	35	28	20
2	27	22	15	53	44	31
2 1/2	38	31	22	75	61	43
3	49	40	29	97	80	56
3 1/2	59	49	34	118	97	68

7-5

SAFE LOADS FOR SHACKLES

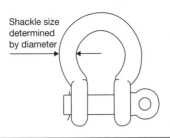

Shackle size determined by diameter

Size (in.)	Safe Load at 90° (tons)
$\frac{1}{4}$	$\frac{1}{3}$
$\frac{5}{16}$	$\frac{1}{2}$
$\frac{3}{8}$	$\frac{3}{4}$
$\frac{7}{16}$	1
$\frac{1}{2}$	$1\frac{1}{2}$
$\frac{5}{8}$	2
$\frac{3}{4}$	3
$\frac{7}{8}$	4
1	$5\frac{1}{2}$
$1\frac{1}{8}$	$6\frac{1}{2}$
$1\frac{1}{4}$	8
$1\frac{3}{8}$	10
$1\frac{1}{2}$	12
$1\frac{3}{4}$	16
2	21
$2\frac{1}{4}$	27
$2\frac{1}{2}$	34
$2\frac{3}{4}$	40
3	50

ROPE CHARACTERISTICS

Rope Diameter (in.)	Safe Load Ratio	Nylon		Polypropylene		Manila	
		Break Lbs.	Lbs./ 100 Feet	Break Lbs.	Lbs./ 100 Feet	Break Lbs.	Lbs./ 100 Feet
3/16	10:1	1000	1.0	800	0.7	406	1.5
1/4	10:1	1650	1.5	1250	1.2	540	2.0
5/16	10:1	2550	2.5	1900	1.8	900	2.9
3/8	10:1	3700	3.5	2700	2.8	1220	4.1
7/16	10:1	5000	5.0	3500	3.8	1580	5.3
1/2	9:1	6400	6.5	4200	4.7	2380	7.5
9/16	8:1	8000	8.3	5100	6.1	3100	10.4
5/8	8:1	10400	10.5	6200	7.5	3960	13.3
3/4	7:1	14200	14.5	8500	10.7	4860	16.7
13/16	7:1	17000	17.0	9900	12.7	5850	19.5
7/8	7:1	20000	20.0	11500	15.0	6950	22.4
1	7:1	25000	26.4	14000	18.0	8100	27.0
1 1/16	7:1	28800	29.0	16000	20.4	9450	31.2
1 1/8	7:1	33000	34.0	18300	23.8	10800	36.0
1 1/4	7:1	37500	40.0	21000	27.0	12200	41.6
1 5/16	7:1	43000	45.0	23500	30.4	13500	47.8
1 1/2	7:1	53000	55.0	29700	38.4	16700	60.0
1 5/8	7:1	65000	66.5	36000	47.6	20200	74.5
1 3/4	7:1	78000	83.0	43000	59.0	23800	89.5
2	7:1	92000	95.0	52000	69.0	28000	108
2 1/8	7:1	106000	109	61000	80.0		
2 1/4	6:1	125000	129	69000	92.0		
2 1/2	6:1	140000	149	80000	107		
2 5/8	6:1	162000	168	90000	120		
2 7/8	6:1	180000	189	101000	137		
3	6:1	200000	210	114000	153		
3 1/4	6:1	250000	264	137000	190		
3 1/2	6:1	300000	312	162000	232		
4	6:1	360000	380	190000	276		

Lbs/foot = Rope weight per linear foot
Break Lbs = Tensile strength
Safe Load Ratio = Break strength to safe load
Example: 7/16" nylon rope break strength = 5000 lbs.
5000/10 = 500 lbs. safe working load
Note: increased temperatures decrease rope strength

MANILA ROPE SLINGS – STRAIGHT LEG

Rope Diameter (in.)	Straight 2 Leg	Straight 4 Leg	Choker 2 Leg	60° Choker 4 Leg	45° Choker 4 Leg	30° Choker 4 Leg
			CAPACITY IN TONS			
1/2	1/2	1	1/3	2/3	1/2	1/3
3/4	3/4	1 1/2	3/4	1 1/4	1	3/4
1	1 1/2	3	1 1/4	2	1 1/2	1 1/4
1 1/2	3	6	2	4	3	2
2	5	10	4	7	6	4
2 1/2	7	15	6	10	8	6
3	10	20	8	14	12	8
3 1/2	14	29	11	20	16	11
4	17	34	13	23	19	13

MANILA ROPE SLINGS – BASKET

Rope Diameter (in.)	60° Basket 6 Leg	45° Basket 6 Leg	60° Basket 6 Leg	30° Basket 4 Leg	45° Basket 4 Leg	60° Basket 4 Leg
	30° Basket 6 Leg	45° Basket 6 Leg	60° Basket 6 Leg	30° Basket 4 Leg	45° Basket 4 Leg	60° Basket 4 Leg
	CAPACITY IN TONS					
½	⅔	¾	1	½	⅔	¾
¾	1	2	2¼	¾	1	1½
1	2	3	4	1½	2	2½
1½	4	6	8	3	4	5
2	7	11	13	5	7	9
2½	11	16	19	7	11	13
3	15	22	27	10	15	18
3½	22	31	38	15	21	25
4	25	36	44	17	24	30

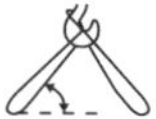

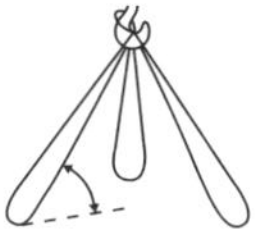

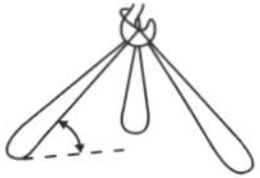

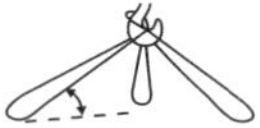

7-9

STANDARD HAND SIGNALS FOR CONTROLLING CRANE OPERATIONS

HOIST. Forearm vertical, forefinger pointing up, move hand in small horizontal circles.

LOWER. Arm extended downward, forefinger pointing down, move hand in small horizontal circles.

USE MAIN HOIST. Tap fist on head; then use regular signals.

USE WHIPLINE. Tap elbow with one hand; then use regular signals.

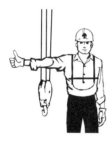

RAISE BOOM. Arm extended, fingers closed, thumb pointing upward.

LOWER BOOM. Arm extended, fingers closed, thumb pointing down.

STANDARD HAND SIGNALS FOR
CONTROLLING CRANE OPERATIONS *(cont.)*

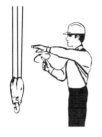

MOVE SLOWLY. One hand gives motion signal, other hand motionless in front of hand giving the motion signal.

RAISE THE BOOM AND LOWER THE LOAD. Arm extended, thumb pointing up, flex fingers in and out.

LOWER THE BOOM AND RAISE THE LOAD. Arm extended, thumb pointing down, flex fingers in and out.

STANDARD HAND SIGNALS FOR
CONTROLLING CRANE OPERATIONS *(cont.)*

SWING. Arm extended, point with finger in direction of swing.

STOP. Arm extended, palm down, hold.

EMERGENCY STOP. Arm extended, palm down, move hand rapidly right and left.

TRAVEL. Arm extended forward, hand open and slightly raised, pushing motion in direction of travel.

EXTEND BOOM. Both fists in front of body with thumbs pointing outward.

RETRACT BOOM. Both fists in front of body with thumbs pointing toward each other.

EYE AND FACE PROTECTORS

1. GOGGLES, Flexible Fitting – Regular Ventilation
2. GOGGLES, Flexible Fitting – Hooded Ventilation
3. GOGGLES, Cushioned Fitting – Rigid Body
4. SPECTACLES, Metal Frame – with Sideshields[1]
5. SPECTACLES, Plastic Frame – with Sideshields
6. SPECTACLES, Metal-Plastic Frame – with Sideshields[1]
7. WELDING GOGGLES, Eyecup Type – Tinted Lenses
7A. CHIPPING GOGGLES, Eyecup Type – Clear Safety Lenses
8. WELDING GOGGLES, Coverspec Type – Tinted Lenses
8A. CHIPPING GOGGLES, Coverspec Type – Clear Safety Lenses
9. WELDING GOGGLES, Coverspec Type – Tinted Plate Lens
10. FACE SHIELD (Available with Plastic or Mesh Window)
11. WELDING HELMETS

[1]Non-side shield spectacles are available for limited hazard use requiring only frontal protection.

7-15

EYE PROTECTION RECOMMENDATIONS

Operation	Hazards	Recommended protectors
Acetylene-Burning, Acetylene-Cutting, Acetylene-Welding	Sparks, harmful rays, molten metal, flying particles	7, 8, 9
Chemical Handling	Splash, acid burns, fumes	2, 10 (For severe exposure add 10 over 2)
Chipping	Flying particles	1, 3, 4, 5, 6, 7A, 8A
Electric (arc) welding	Sparks, intense rays, molten metal	9, 11, (11 in combination with 4, 5, 6, in tinted lenses advisable)
Furnace operations	Glare, heat, molten metal	7, 8, 9 (For severe exposure add 10)
Grinding-Light	Flying particles	1, 3, 4, 5, 6, 10
Grinding-Heavy	Flying particles	1, 3, 7A, 8A (For severe exposure add 10)
Laboratory	Chemical splash, glass breakage	2 (10 when in combination with 4, 5, 6)
Machining	Flying particles	1, 3, 4, 5, 6, 10
Molten metals	Heat, glare, sparks, splash	7, 8, (10 in combination with 4, 5, 6, in tinted lenses)
Spot welding	Flying particles, sparks	1, 3, 4, 5, 6, 10

FILTER LENS SHADE NUMBERS FOR PROTECTION AGAINST RADIANT ENERGY

Welding operation	Shade number
Gas-shielded arc welding (nonferrous) 1/16, 3/32, 1/8, 5/32 inch diameter electrodes	11
Gas-shielded arc welding (ferrous) 1/16, 3/32, 1/8, 5/32 inch diameter electrodes	12
Shielded metal-arc welding 3/32, 1/8, 5/32 inch diameter electrodes	10
Shielded metal-arc welding 3/16, 7/32, 1/4 inch diameter electrodes	12
Shielded metal-arc welding 5/16, 3/8 inch diameter electrodes	14
Atomic hydrogen welding	10 to 14
Carbon-arc welding	14
Soldering	2
Torch brazing	3 or 4
Light cutting, up to 1 inch	3 or 4
Medium cutting, 1 inch to 6 inches	4 or 5
Heavy cutting, over 6 inches	5 or 6
Gas welding (light), up to 1/8 inch	4 or 5
Gas welding (medium), 1/8 inch to 1/2 inch	5 or 6
Gas welding (heavy), over 1/2 inch	6 or 8

SELECTING RESPIRATORS BY HAZARD

Hazard	Respirator
Oxygen deficiency	Self-contained breathing apparatus. Hose mask with blower. Combination air-line respirator with auxiliary self-contained air supply or an air-storage receiver with alarm.
Gas and vapor contaminants immediately dangerous to life and health	Self-contained breathing apparatus. Hose mask with blower. Air-purifying full facepiece respirator (for escape only). Combination air-line respirator with auxiliary self-contained air supply or an air-storage receiver with alarm.
Not immediately dangerous to life and health	Air-line respirator. Hose mask without blower. Air-purifying, half-mask, or mouth-piece respirator with chemical cartridge.
Particulate contaminants immediately dangerous to life and health	Self-contained breathing apparatus. Hose mask with blower. Air-purifying, full facepiece respirator with appropriate filter. Self-rescue mouthpiece respirator (for escape only). Combination air-line respirator with auxiliary self-contained air supply or an air-storage receiver with alarm.
Not immediately dangerous to life and health	Air-purifying, half-mask, or mouthpiece respirator with filter pad or cartridge. Air-line respirator. Air-line abrasive-blasting respirator. Hose-mask without blower.

Immediately dangerous to life and health is a condition that either poses an immediate threat of severe exposure to contaminants (radioactive materials) or which are likely to have adverse delayed effects on health.

SELECTING RESPIRATORS BY HAZARD *(cont.)*

Hazard	Respirator
Combination gas, vapor, and particulate contaminants immediately dangerous to life and health	Self-contained breathing apparatus. Hose mask with blower. Air-purifying, full facepiece respirator with chemical canister and appropriate filter (gas mask with filter). Self-rescue mouthpiece respirator (for escape only). Combination air-line respirator with auxiliary self-contained air-supply or an air-storage receiver with alarm.
Not immediately dangerous to life and health	Air-line respirator. Hose mask without blower. Air-purifying, half-mask, or mouthpiece respirator with chemical cartridge and appropriate filter.

Immediately dangerous to life and health is a condition that either poses an immediate threat of severe exposure to contaminants (radioactive materials) or which are likely to have adverse delayed effects on health.

PERMISSIBLE NOISE EXPOSURES

Duration per Day (Hours)	Sound Level in dBA (Slow Response)
8	90
6	92
4	95
3	97
2	100
1½	102
1	105
½	110
¼ or less	115

MINIMUM ILLUMINATION INTENSITIES IN FOOT-CANDLES

Foot-candles	Area of Operation
3	General construction areas, concrete placement, excavation and waste areas, access ways, active storage areas, loading platforms, refueling, and field maintenance areas.
5	General construction area lighting.
5	Indoors: warehouses, corridors, hallways, and exitways.
5	Tunnels, shafts, and general underground work areas. (Exception: minimum of 10 foot-candles is required at tunnel heading during drilling, mucking, and scaling. Bureau of Mines approved cap lights shall be acceptable for use in the tunnel heading.)
10	General construction plant and shops. (e.g., batch plants, screening plants, mechanical and electrical equipment rooms, carpenter shops, rigging lofts and active store rooms, mess halls, and indoor toilets and workrooms)
30	First aid stations, infirmaries, and offices.

TYPES OF FIRE EXTINGUISHERS

TYPE A: To extinguish fires involving trash, cloth, paper, and other wood- or pulp-based materials. The flames are put out by water-based ingredients or dry chemicals.

TYPE B: To extinguish fires involving greases, paints, solvents, gas, and other petroleum-based liquids. The flames are put out by cutting off oxygen and stopping the release of flammable vapors. Dry chemicals, foams, and halon are used.

TYPE C: To extinguish fires involving electricity. The combustion is put out the same way as with a type B extinguisher but, most importantly, the chemical in a type C <u>MUST</u> be non-conductive to electricity in order to be safe and effective.

TYPE D: To extinguish fires involving combustible metals. Please be advised to obtain important information from your local fire department on the requirements for type D fire extinguishers for your area.

Any combination of letters indicate that an extinguisher will put out more than one type of fire. Type BC will put out two types of fires. The size of the fire to be extinguished is shown by a number in front of the letter such as 100A. For example:

Class 1A will extinguish 25 burning sticks 40 inches long.

Class 1B will extinguish a paint thinner fire 2.5 square feet in size.

Class 100B will put out a fire 100 times larger than type 1B.

Here are some basic guidelines to follow:

• By using a type ABC you will cover most basic fires.

• Use fire extinguishers with a gauge and that are constructed with metal. Also note if the unit is U.L. approved.

• Utilize more than one extinguisher and be sure that each unit is mounted in a clearly visible and accessible manner.

• After purchasing any fire extinguisher always review the basic instructions for its intended use. Never deviate from the manufacturer's guidelines. Following this simple procedure could end up saving lives.

EXTENSION CORD SIZES FOR PORTABLE TOOLS

Cord Length (Feet)	Full-Load Rating of Tool in Amperes at 115 Volts					
	0 to 2.0	2.1 to 3.4	3.5 to 5.0	5.1 to 7.0	7.1 to 12.0	12.1 to 16.0
			Wire Size (AWG)			
25	18	18	18	16	14	14
50	18	18	18	16	14	12
75	18	18	16	14	12	10
100	18	16	14	12	10	8
200	16	14	12	10	8	6
300	14	12	10	8	6	4
400	12	10	8	6	4	4
500	12	10	8	6	4	2
600	10	8	6	4	2	2
800	10	8	6	4	2	1
1000	8	6	4	2	1	0

CHAPTER 8
Materials and Tools

		Weight of Wrench (lbs.)		
Pipe Size (In.)	Wrench Length (In.)	Steel Straight and End Wrench	Aluminum Straight Wrench	Aluminum End Wrench
3/4	6	.50	–	–
1	8	.75	–	–
1½	10	1.75	1.00	–
2	12	2.75	–	–
2	14	3.50	2.34	1.75
2½	18	5.75	3.67	3.50
3	24	9.75	6.00	5.75
5	36	19.00	11.00	–
6	48	34.25	18.50	–
8	60	51.25	–	–

WRENCHES FOR USE WITH VARIOUS PIPE SIZES

BAND SAW TEETH PER INCH AND SPEED

Type of Material To Be Cut	Size of Material			
	½"-1"	1"-2"	½"-1"	1"-2"
	Teeth-per-inch		Speed (fpm)	
Steels				
Angle Iron	14	14	190	175
Armor plate	14	12	100	75
Cast Iron	12	10	200	185
Cast steels	14	12	150	75
Graphic steel	14	12	150	125
High-speed steel	14	10	100	75
I-beams and channels	14	14	250	200
Pipe	14	12	250	225
Stainless steel	12	10	60	50
Tubing (thinwall)	14	14	250	200
Non-ferrous Metals				
Aluminum (all types)	8	6	250	250
Beryllium	10	8	175	150
Brass	8	8	250	250
Bronze (cast)	10	8	185	125
Bronze (rolled)	12	10	175	125
Copper	10	8	250	225
Magnesium	8	8	250	250

RECOMMENDED DRILLING SPEEDS IN RPM

Material	Bit Sizes (in.)	RPM Speed Range		
Glass	Special Metal Tube Drilling	700		
Plastics	7/16 and larger	500	–	1000
	3/8	1500	–	2000
	5/16	2000	–	2500
	1/4	3000	–	3500
	3/16	3500	–	4000
	1/8	5000	–	6000
	1/16 and smaller	6000	–	6500
Woods	1 and larger	700	–	2000
	3/4 to 1	2000	–	2300
	1/2 to 3/4	2300	–	3100
	1/4 to 1/2	3100	–	3800
	1/4 and smaller	3800	–	4000
	carving / routing	4000	–	6000
Soft Metals	7/16 and larger	1500	–	2500
	3/8	3000	–	3500
	5/16	3500	–	4000
	1/4	4500	–	5000
	3/16	5000	–	6000
	1/8	6000	–	6500
	1/16 and smaller	6000	–	6500
Steel	7/16 and larger	500	–	1000
	3/8	1000	–	1500
	5/16	1000	–	1500
	1/4	1500	–	2000
	3/16	2000	–	2500
	1/8	3000	–	4000
	1/16 and smaller	5000	–	6500
Cast Iron	7/16 and larger	1000	–	1500
	3/8	1500	–	2000
	5/16	1500	–	2000
	1/4	2000	–	2500
	3/16	2500	–	3000
	1/8	3500	–	4500
	1/16 and smaller	6000	–	6500

STANDARD TAPS AND DIES IN INCHES

Thread Size	Coarse			Fine		
	Drill Size	Threads Per Inch	Decimal Size	Drill Size	Threads Per Inch	Decimal Size
4	3	4	3.75	–	–	–
$3\frac{3}{4}$	3	4	3.5	–	–	–
$3\frac{1}{2}$	3	4	3.25	–	–	–
$3\frac{1}{4}$	3	4	3.0	–	–	–
3	2	4	2.75	–	–	–
$2\frac{3}{4}$	2	4	2.5	–	–	–
$2\frac{1}{2}$	2	4	2.25	–	–	–
$2\frac{1}{4}$	2	4.5	2.0313	–	–	–
2	1	4.5	1.7813	–	–	–
$1\frac{3}{4}$	1	2	1.5469	–	–	–
$1\frac{1}{2}$	1	6	1.3281	$1\frac{27}{64}$	12	1.4219
$1\frac{3}{8}$	1	6	1.2188	$1\frac{19}{64}$	12	1.2969
$1\frac{1}{4}$	1	7	1.1094	$1\frac{11}{64}$	12	1.1719
$1\frac{1}{8}$	$\frac{63}{64}$	7	.9844	$1\frac{3}{64}$	12	1.0469
1	$\frac{7}{8}$	8	.8750	$\frac{15}{16}$	14	.9375
$\frac{7}{8}$	$\frac{49}{64}$	9	.7656	$\frac{13}{16}$	14	.8125
$\frac{3}{4}$	$\frac{21}{32}$	10	.6563	$\frac{11}{16}$	16	.6875
$\frac{5}{8}$	$\frac{17}{32}$	11	.5313	$\frac{37}{64}$	18	.5781
$\frac{9}{16}$	$\frac{31}{64}$	12	.4844	$\frac{33}{64}$	18	.5156
$\frac{1}{2}$	$\frac{27}{64}$	13	.4219	$\frac{29}{64}$	20	.4531
$\frac{7}{16}$	U	14	.368	$\frac{25}{64}$	20	.3906
$\frac{3}{8}$	$\frac{5}{16}$	16	.3125	Q	24	.332
$\frac{5}{16}$	F	18	.2570	I	24	.272
$\frac{1}{4}$	#7	20	.201	#3	28	.213
#12	#16	24	.177	#14	28	.182
#10	#25	24	.1495	#21	32	.159
$\frac{3}{16}$	#26	24	.147	#22	32	.157
#8	#29	32	.136	#29	36	.136
#6	#36	32	.1065	#33	40	.113
#5	#38	40	.1015	#37	44	.104
$\frac{1}{8}$	$\frac{3}{32}$	32	.0938	#38	40	.1015
#4	#43	40	.089	#42	48	.0935
#3	#47	48	.0785	#45	56	.082
#2	#50	56	.07	#50	64	.07
#1	#53	64	.0595	#53	72	.0595
#0	–	–	–	$\frac{3}{64}$	80	.0469

TAPS AND DIES – METRIC CONVERSIONS

Thread Pitch (mm)	Fine Thread Size		Tap Drill Size	
	(in.)	(mm)	(in.)	(mm)
4.5	1.6535	42	1.4567	37.0
4.0	1.5748	40	1.4173	36.0
4.0	1.5354	39	1.3779	35.0
4.0	1.4961	38	1.3386	34.0
4.0	1.4173	36	1.2598	32.0
3.5	1.3386	34	1.2008	30.5
3.5	1.2992	33	1.1614	29.5
3.5	1.2598	32	1.1220	28.5
3.5	1.1811	30	1.0433	26.5
3.0	1.1024	28	.9842	25.0
3.0	1.0630	27	.9449	24.0
3.0	1.0236	26	.9055	23.0
3.0	.9449	24	.8268	21.0
2.5	.8771	22	.7677	19.5
2.5	.7974	20	.6890	17.5
2.5	.7087	18	.6102	15.5
2.0	.6299	16	.5118	14.0
2.0	.5512	14	.4724	12.0
1.75	.4624	12	.4134	10.5
1.50	.4624	12	.4134	10.5
1.50	.3937	11	.3780	9.6
1.50	.3937	10	.3386	8.6
1.25	.3543	9	.3071	7.8
1.25	.3150	8	.2677	6.8
1.0	.2856	7	.2362	6.0
1.0	.2362	6	.1968	5.0
.90	.2165	5.5	.1811	4.6
.80	.1968	5	.1653	4.2
.75	.1772	4.5	.1476	3.75
.70	.1575	4	.1299	3.3
.75	.1575	4	.1279	3.25
.60	.1378	3.5	.1142	2.9
.60	.1181	3	.0945	2.4
.50	.1181	3	.0984	2.5
.45	.1124	2.6	.0827	2.1
.45	.0984	2.5	.0787	2.0
.40	.0895	2.3	.0748	1.9
.40	.0787	2	.0630	1.6
.45	.0787	2	.0590	1.5
.35	.0590	1.5	.0433	1.1

TYPES OF SOLDERING FLUX

To Solder	Use
Cast iron	Cuprous oxide
Galvanized iron, galvanized, steel, tin, zinc	Hydrochloric acid
Pewter and lead	Organic
Brass, copper, gold, iron, silver, steel	Borax
Brass, bronze, cadmium, copper, lead, silver	Resin
Brass, copper, gun metal, iron, nickel, tin, zinc	Ammonia chloride
Bismuth, brass, copper, gold, silver, tin	Zinc chloride
Silver	Sterling
Pewter and lead	Tallow
Stainless steel	Stainless steel (only)

HARD SOLDER ALLOYS

To hard solder	Copper %	Gold %	Silver %	Zinc %
Gold	22	67	11	–
Silver	20	–	70	10
Hard brass	45	–	–	55
Soft brass	22	–	–	78
Copper	50	–	–	50
Cast iron	55	–	–	45
Steel and iron	64	–	–	36

SOFT SOLDER ALLOYS

To soft solder	Lead %	Tin %	Zinc %	Bism %	Other %
Gold	33	67	–	–	–
Silver	33	67	–	–	–
Brass	34	66	–	–	–
Copper	40	60	–	–	–
Steel and iron	50	50	–	–	–
Galvanized steel	42	58	–	–	–
Tinned steel	36	64	–	–	–
Zinc	45	55	–	–	–
Block Tin	1	99	–	–	–
Lead	67	33	–	–	–
Gun metal	37	63	–	–	–
Pewter	25	25	–	50	–
Bismuth	33	33	–	34	–
Aluminum	–	70	25	–	5

PROPERTIES OF WELDING GASES

Gas Type	Gas Characteristics	Tank Sizes (cu. ft.)
Acetylene	C_2H_2, explosive gas, flammable, garlic - like odor, colorless, dangerous if used in pressures over 15 psig (30 psig absolute)	10, 40, 75 100, 300
Argon	Ar, non-explosive inert gas, tasteless, odorless, colorless	131, 330 4754 (Liquid)
Carbon Dioxide	CO_2, Non-explosive inert gas, tasteless, odorless, colorless (in large quantities is toxic)	20 lbs., 50 lbs.
Helium	He, Non-explosive inert gas, tasteless, odorless, colorless	221
Hydrogen	H_2, explosive gas, tasteless, odorless, colorless	191
Nitrogen	N_2, Non-explosive inert gas, tasteless, odorless, colorless	20, 40, 80 113, 225
Oxygen	O_2, Non-explosive gas, tasteless, odorless, colorless, supports combustion	20, 40, 80 122, 244 4500 (liquid)

WELDING RODS – 36" LONG

Rod Size (In.)	Number of Rods Per Pound			
	Aluminum	Brass	Cast Iron	Steel
3/8	–	1.0	.25	1.0
5/16	–	–	.50	1.33
1/4	6.0	2.0	2.25	2.0
3/16	9.0	3.0	5.50	3.5
5/32	–	–	–	5.0
1/8	23.0	7.0	–	8.0
3/32	41.0	13.0	–	14.0
1/16	91.0	29.0	–	31.0

METALWORKING LUBRICANTS

Material	Threading	Drilling	Lathing
Machine Steel	Dissolvable Oil Mineral Oil Lard Oil	Dissolvable Oil Sulpherized Oil Mineral Lard Oil	Dissolvable Oil
Tool Steel	Lard Oil Sulpherized Oil	Dissolvable Oil Sulpherized Oil	Dissolvable Oil
Steel Alloys	Lard Oil Sulpherized Oil	Dissolvable Oil Sulpherized Oil Mineral Lard Oil	Dissolvable Oil
Cast Iron	Sulpherized Oil Dry Mineral Lard Oil	Dissolvable Oil Dry Air Jet	Dissolvable Oil Dry
Malleable Iron	Soda Water Lard Oil	Soda Water Dry	Soda Water Dissolvable Oil
Monel Metal	Lard Oil	Dissolvable Oil Lard Oil	Dissolvable Oil
Aluminum	Kerosene Dissolvable Oil Lard Oil	Kerosene Dissolvable Oil	Dissolvable Oil
Brass	Dissolvable Oil Lard Oil	Kerosene Dissolvable Oil Dry	Dissolvable Oil
Bronze	Dissolvable Oil Lard Oil	Dissolvable Oil Dry Mineral Oil Lard Oil	Dissolvable Oil
Copper	Dissolvable Oil Lard Oil	Kerosene Dissolvable Oil Dry Mineral Lard Oil	Dissolvable Oil

TORQUE LUBRICATION EFFECTS IN FOOT-POUNDS

Lubricant	5/16" – 18 Thread	1/2" – 13 Thread	Torque Decrease
Graphite	13	62	49 – 55%
Mily Film	14	66	45 – 52%
White Grease	16	79	35 – 45%
SAE 30	16	79	35 – 45%
SAE40	17	83	31 – 41%
SAE 20	18	87	28 – 38%
Plated	19	90	26 – 34%
No Lube	29	121	0%

TIGHTENING TORQUE IN POUND-FEET-SCREW FIT

Wire Size, AWG/kcmil	Driver	Bolt	Other
18-16	1.67	6.25	4.2
14-8	1.67	6.25	6.125
6-4	3.0	12.5	8.0
3-1	3.2	21.00	10.40
0-2/0	4.22	29	12.5
3/0-200	–	37.5	17.0
250-300	–	50.0	21.0

SCREW TORQUES

Screw Size, Inches Across, Hex Flats	Torque, Pound-Feet
1/8	4.2
5/32	8.3
3/16	15.0
7/32	23.25
1/4	42.0

SHEET METAL SCREW CHARACTERISTICS

Screw Size #	Screw Diameter (in.)	Diameter of Pierced Hole (in.)	Hole Size #	Metal Gauge #
#4	.112	.086	#44	28
		.086	#44	26
		.093	#42	24
		.098	#42	22
		.100	#40	20
#6	.138	.111	#39	28
		.111	#39	26
		.111	#39	24
		.111	#38	22
		.111	#36	20
#7	.155	.121	#37	28
		.121	#37	26
		.121	#35	24
		.121	#33	22
		.121	#32	20
		–	#31	18
#8	.165	.137	#33	26
		.137	#33	24
		.137	#32	22
		.137	#31	20
		–	#30	18
#10	.191	.158	#30	26
		.158	#30	24
		.158	#30	22
		.158	#29	20
		.158	#25	18
#12	.218	–	#26	24
		.185	#25	22
		.185	#24	20
		.185	#22	18
#14	.251	–	#15	24
		.212	#12	22
		.212	#11	20
		.212	#9	18

STANDARD WOOD SCREW CHARACTERISTICS

Screw Size #	Wood Screw Lengths (in.)	Pilot Hole		Shank Hole	
		Softwood Bit #	Hardwood Bit #	Clearance Bit #	Hole Diameter (in.)
0	1/4	75	66	52	.060
1	1/4 to 3/8	71	57	47	.073
2	1/4 to 1/2	65	54	42	.086
3	1/4 to 5/8	58	53	37	.099
4	3/8 to 3/4	55	51	32	.112
5	3/8 to 3/4	53	47	30	.125
6	3/8 to 1 1/2	52	44	27	.138
7	3/8 to 1 1/2	51	39	22	.151
8	1/2 to 2	48	35	18	.164
9	5/8 to 2 1/4	45	33	14	.177
10	5/8 to 2 1/2	43	31	10	.190
11	3/4 to 3	40	29	4	.203
12	7/8 to 3 1/2	38	25	2	.216
14	1 to 4 1/2	32	14	D	.242
16	1 1/4 to 5 1/2	29	10	I	.268
18	1 1/2 to 6	26	6	N	.294
20	1 3/4 to 6	19	3	P	.320
24	3 1/2 to 6	15	D	V	.372

ALLEN HEAD AND MACHINE SCREW
BOLT AND TORQUE CHARACTERISTICS

Number of Threads Per Inch	Allen Head and Mach. Screw Bolt Size (in.)	Allen Head Case H Steel 160,000 psi	Mach. Screw Yellow Brass 60,000 psi	Mach. Screw Silicone Bronze 70,000 psi
		Torque in Foot-Pounds or Inch-Pounds		
4.5	2	8800	–	–
5	1¾	6100	–	–
6	1½	3450	655	595
6	1⅜	2850	–	–
7	1¼	2130	450	400
7	1⅛	1520	365	325
8	1	970	250	215
9	⅞	640	180	160
10	¾	400	117	104
11	⅝	250	88	78
12	⁹⁄₁₆	180	53	49
13	½	125	41	37
14	⁷⁄₁₆	84	30	27
16	⅜	54	20	17
18	⁵⁄₁₆	33	125 in#	110 in#
20	¼	16	70 in#	65 in#
24	#10	60	22 in#	20 in#
32	#8	46	19 in#	16 in#
32	#6	21	10 in#	8 in#
40	#5	–	7.2 in#	6.4 in#
40	#4	–	4.9 in#	4.4 in#
48	#3	–	3.7 in#	3.3 in#
56	#2	–	2.3 in#	2 in#

For fine thread bolts, increase by 9%.

HEX HEAD BOLT AND TORQUE CHARACTERISTICS

BOLT MAKE-UP IS STEEL WITH COARSE THREADS

Number of Threads Per Inch	Hex Head Bolt Size (in.)	SAE 0-1-2 74,000 psi	 SAE Grade 3 100,000 psi	 SAE Grade 5 120,000 psi
		Torque in Foot-Pounds		
4.5	2	2750	5427	4550
5	1¾	1900	3436	3150
6	1½	1100	1943	1775
6	1⅜	900	1624	1500
7	1¼	675	1211	1105
7	1⅛	480	872	794
8	1	310	551	587
9	⅞	206	372	382
10	¾	155	234	257
11	⅝	96	145	154
12	9⁄16	69	103	114
13	½	47	69	78
14	7⁄16	32	47	54
16	⅜	20	30	33
18	5⁄16	12	17	19
20	¼	6	9	10

For fine thread bolts, increase by 9%.

HEX HEAD BOLT AND TORQUE CHARACTERISTICS (cont.)

BOLT MAKE-UP IS STEEL WITH COARSE THREADS

Number of Threads Per Inch	Hex Head Bolt Size (in.)	SAE Grade 6 133,000 psi	SAE Grade 7 133,000 psi	SAE Grade 8 150,000 psi
		Torque in Foot-Pounds		
4.5	2	7491	7500	8200
5	1¾	5189	5300	5650
6	1½	2913	3000	3200
6	1⅜	2434	2500	2650
7	1¼	1815	1825	1975
7	1⅛	1304	1325	1430
8	1	825	840	700
9	⅞	550	570	600
10	¾	350	360	380
11	⅝	209	215	230
12	⁹⁄₁₆	150	154	169
13	½	106	110	119
14	⁷⁄₁₆	69	71	78
16	⅜	43	44	47
18	⁵⁄₁₆	24	25	29
20	¼	12.5	13	14

For fine thread bolts, increase by 9%.
For special alloy bolts, obtain torque rating from the manufacturer.

WHITWORTH HEX HEAD BOLT AND TORQUE CHARACTERISTICS

BOLT MAKE-UP IS STEEL WITH COARSE THREADS

Number of Threads Per Inch	Whitworth Type Hex Head Bolt Size (in.)	Grades A & B 62,720 psi	Grade S 112,000 psi	Grade T 123,200 psi	Grade V 145,600 psi
		⬡	⬡	⬡	⬡
			Torque in Foot-Pounds		
8	1	276	497	611	693
9	7/8	186	322	407	459
11	3/4	118	213	259	287
11	5/8	73	128	155	175
12	9/16	52	94	111	128
12	1/2	36	64	79	89
14	7/16	24	43	51	58
16	3/8	15	27	31	36
18	5/16	9	15	18	21
20	1/4	5	7	9	10

For fine thread bolts, increase by 9%.

8-15

METRIC HEX HEAD BOLT AND TORQUE CHARACTERISTICS

BOLT MAKE-UP IS STEEL WITH COARSE THREADS

Thread Pitch (mm)	Bolt Size (mm)	Standard 5D 71,160 psi	Standard 8G 113,800 psi	Standard 10K 142,000 psi	Standard 12K 170,674 psi
			Torque in Foot-Pounds		
3.0	24	261	419	570	689
2.5	22	182	284	394	464
2.0	18	111	182	236	183
2.0	16	83	132	175	208
1.25	14	55	89	117	137
1.25	12	34	54	70	86
1.25	10	19	31	40	49
1.0	8	10	16	22	27
1.0	6	5	6	8	10

For fine thread bolts, increase by 9%.

PULLEY AND GEAR FORMULAS

For single reduction or increase of speed by means of belting where the speed at which each shaft should run is known, and one pulley is in place:

Multiply the diameter of the pulley which you have by the number of revolutions per minute that its shaft makes; divide this product by the speed in revolutions per minute at which the second shaft should run. The result is the diameter of pulley to use.

Where both shafts with pulleys are in operation and the speed of one is known:

Multiply the speed of the shaft by diameter of its pulley and divide this product by diameter of pulley on the other shaft. The result is the speed of the second shaft.

Where a countershaft is used, to obtain size of main driving or driven pulley, or speed of main driving or driven shaft, it is necessary to calculate, as above, between the known end of the transmission and the countershaft, then repeat this calculation between the countershaft and the unknown end.

A set of gears of the same pitch transmits speeds in proportion to the number of teeth they contain. Count the number of teeth in the gear wheel and use this quantity instead of the diameter of pulley, mentioned above, to obtain number of teeth cut in unknown gear, or speed of second shaft.

Formulas For Finding Pulley Sizes:

$$d = \frac{D \times S}{s'} \qquad D = \frac{d \times s'}{S}$$

d = diameter of driven pulley

D = diameter of driving pulley

s' = number of revolutions per minute of driven pulley

S = number of revolutions per minute of driving pulley

PULLEY AND GEAR FORMULAS (cont.)

Formulas For Finding Gear Sizes:

$$n = \frac{N \times S}{s'} \qquad N = \frac{n \times s'}{S}$$

n = number of teeth in pinion (driving gear)

N = number of teeth in gear (driven gear)

s' = number of revolutions per minute of gear

S = number of revolutions per minute of pinion

Formula To Determine Shaft Diameter:

$$\text{diameter of shaft in inches} = \sqrt[3]{\frac{K \times HP}{RPM}}$$

HP = the horsepower to be transmitted

RPM = speed of shaft

K = factor which varies from 50 to 125 depending on type of shaft and distance between supporting bearings.

For line shaft having bearings 8 feet apart:

K = 90 for turned shafting

K = 70 for cold-rolled shafting

Formula To Determine Belt Length:

$$\text{length of belt} = \frac{3.14\,(D+d)}{2} + 2\left(\sqrt{X^2 + \left(\frac{D-d}{2}\right)^2}\right)$$

D = diameter of large pulley

d = diameter of small pulley

X = distance between centers of shafting

STANDARD "V" BELT LENGTHS IN INCHES

A BELTS		
Standard Belt No.	Pitch Length	Outside Length
A26	27.3	28.0
A31	32.3	33.0
A35	36.3	37.0
A38	39.3	40.0
A42	43.3	44.0
A46	47.3	48.0
A51	52.3	53.0
A55	56.3	57.0
A60	61.3	62.0
A68	69.3	70.0
A75	76.3	77.0
A80	81.3	82.0
A85	86.3	87.0
A90	91.3	92.0
A96	97.3	98.0
A105	106.3	107.0
A112	113.3	114.0
A120	121.3	122.0
A128	129.3	130.0

B BELTS		
Standard Belt No.	Pitch Length	Outside Length
B35	36.8	38.0
B38	39.8	41.0
B42	43.8	45.0
B46	47.8	49.0
B51	52.8	54.0
B55	56.8	58.0
B60	61.8	63.0
B68	69.8	71.0
B75	76.8	78.0
B81	82.8	84.0
B85	86.8	88.0
B90	91.8	93.0
B97	98.8	100.0
B105	106.8	108.0
B112	113.8	115.0
B120	121.8	123.0
B128	129.8	131.0
B136	137.8	139.0
B144	145.8	147.0
B158	159.8	161.0
B173	174.8	176.0
B180	181.8	183.0
B195	196.8	198.0
B210	211.8	213.0
B240	240.3	241.5
B270	270.3	271.5
B300	300.3	301.5

C BELTS		
Standard Belt No.	Pitch Length	Outside Length
C51	53.9	55.0
C60	62.9	64.0
C68	70.9	81.0
C75	77.9	79.0
C81	83.9	85.0
C85	87.9	89.0
C90	92.9	94.0
C96	98.9	100.0
C105	107.9	109.0
C112	114.9	116.0
C120	122.9	124.0
C128	130.9	132.0
C136	138.9	140.0
C144	146.9	148.0
C158	160.9	162.0
C162	164.9	166.0
C173	175.9	177.0
C180	182.9	184.0
C195	197.9	199.0
C210	212.9	214.0
C240	240.9	242.0
C270	270.9	272.0
C300	300.9	302.0
C360	360.9	362.0
C390	390.9	392.0
C420	420.9	422.0

D BELTS		
Standard Belt No.	Pitch Length	Outside Length
D120	123.3	125.0
D128	131.3	133.0
D144	147.3	149.0
D158	1613	163.0
D162	165.3	167.0
D173	176.3	178.0
D180	183.3	185.0
D195	198.3	200.0
D210	213.3	215.0
D240	240.8	242.0
D270	270.8	272.5
D300	300.8	302.5
D330	330.8	332.5
D360	360.8	362.5
D390	390.8	392.5
D420	420.8	422.5
D480	480.8	482.5
D540	540.8	542.5
D600	600.8	602.5

E BELTS			E BELTS		
Standard Belt No.	Pitch Length	Outside Length	Standard Belt No.	Pitch Length	Outside Length
E180	184.5	187.5	E360	361.0	364.0
E195	199.5	202.5	E390	391.0	394.0
E210	214.5	217.5	E420	421.0	424.0
E240	241.0	244.0	E480	481.0	484.0
E270	271.0	274.0	E540	541.0	544.0
E300	301.0	304.0	E600	601.0	604.0
E330	331.0	334.0			

3V Belts		5V Belts		8V Belts	
3V250	25.0	5V500	50.0	8V1000	100.0
3V265	26.5	5V530	53.0	8V1060	106.0
3V280	28.0	5V560	56.0	8V1120	112.0
3V300	30.0	5V600	60.0	8V1180	118.0
3V315	31.5	5V630	63.0	8V1250	125.0
3V335	33.5	5V670	67.0	8V1320	132.0
3V355	35.5	5V710	71.0	8V1400	140.0
3V375	37.5	5V750	75.0	8V1500	150.0
3V400	40.0	5V800	80.0	8V1600	160.0
3V425	42.5	5V850	85.0	8V1700	170.0
3V450	45.0	5V900	90.0	8V1800	180.0
3V475	47.5	5V950	95.0	8V1900	190.0
3V500	50.0	5V1000	100.0	8V2000	200.0
3V530	53.0	5V1060	106.0	8V2120	212.0
3V560	56.0	5V1120	112.0	8V2240	224.0
3V600	60.0	5V1180	118.0	8V2360	236.0
3V630	63.0	5V1250	125.0	8V2500	250.0
3V670	67.0	5V1320	132.0	8V2650	265.0
3V710	71.0	5V1400	140.0	8V2800	280.0
3V750	75.0	5V1500	150.0	8V3000	300.0
3V800	80.0	5V1600	160.0	8V3150	315.0
3V850	85.0	5V1700	170.0	8V3350	335.0
3V900	90.0	5V1800	180.0	8V3550	355.0
3V950	95.0	5V1900	190.0	8V3750	375.0
3V1000	100.0	5V2000	200.0	8V4000	400.0
3V1060	106.0	5V2120	212.0	8V4250	425.0
3V1120	112.0	5V2240	224.0	8V4500	450.0
3V1180	118.0	5V2360	236.0	8V5000	500.0
3V1250	128.0	5V2500	250.0		
3V1320	132.0	5V2650	265.0		
3V1400	140.0	5V2800	280.0		
		5V3000	300.0		
		5V3150	315.0		
		5V3350	335.0		
		5V3550	355.0		

If the 60-inch "B" section belt shown is made 3/10 of an inch longer, it will be code marked 53 rather than 50. If made 3/10 shorter, it will be marked 47. While both have the belt number B60 they cannot be used in a set because of the difference in length.

TYPICAL CODE MARKING

B60   MANUFACTURER'S NAME   50

NOMINAL
SIZE AND LENGTH

LENGTH
CODE NUMBER

CHAPTER 9
Conversion Factors

COMMONLY USED CONVERSION FACTORS		
Multiply	**By**	**To Obtain**
Acres.	43,560	Square feet
Acre-Feet	43,560	Cubic feet
Amperes per sq. cm.	6.452	Amperes per sq. in.
Amperes per sq. in.	0.1550	Amperes per sq. cm
Ampere-Turns per cm. . . .	2.540	Ampere-turns per in.
Ampere-Turns per in.	0.3937	Ampere-turns per cm
Atmospheres	76.0	Cm of mercury
Atmospheres	29.92	Inches of mercury
Atmospheres	1033.29	Cm of water
Atmospheres	33.90	Feet of water
Atmospheres	101.325	Kilopascals
Atmospheres	101325	Pascals
Atmospheres	14.70	Pounds per sq. in.
British thermal units	252.0	Calories
British thermal units	777.649	Foot pound-force
British thermal units	3.930×10^{-4}	Horsepower-hours
British thermal units	0.2520	Kilogram-calories
British thermal units	107.514	Kilogram-meters
British thermal units	2.931×10^{-4}	Kilowatt-hours
British thermal units	1,054.35	Watt-seconds
B.t.u. per hour	2.931×10^{-4}	Kilowatts
B.t.u. per minute.	0.02358	Horsepower
B.t.u. per minute.	0.01758	Kilowatts
Calories, g	0.003971	Btu
Calories, g	3.086	Foot pound-force

COMMONLY USED CONVERSION FACTORS *(cont.)*

Multiply	By	To Obtain
Calories, kg.	3.968	Btu
Calories, kg.	3.085.96	Foot pound-force
Centimeters	0.3937	Inches
Centimeters	0.03281	Feet
Centimeters	0.01	Meters
Circular mils	5.067×10^{-6}	Square centimeters
Circular mils	0.7854×10^{-6}	Square inches
Circular mils	0.7854	Square mils
Cords	128	Cubic feet
Cubic centimeters	0.06102	Cubic inches
Cubic centimeters	3.5315×10^{-5}	Cubic feet
Cubic centimeters	2.6×10^{-4}	Gallons
Cubic feet.	0.02832	Cubic meters
Cubic feet.	7.481	Gallons
Cubic feet.	28.317	Liters
Cubic inches.	16.3871	Cubic centimeters
Cubic inches.	5.79×10^{-4}	Cubic feet
Cubic inches.	0.00433	Gallons
Cubic meters	35.3147	Cubic feet
Cubic meters	1.308	Cubic yards
Cubic meters	264.172	Gallons
Cubic yards	0.7646	Cubic meters
Cubic yards	201.974	Gallons
Decimeters	10	Centimeters
Decimeters	0.32808	Feet
Decimeters	3.937	Inches
Degrees (angle).	0.00278	Circles
Degrees (angle).	60	Minutes
Degrees (angle).	0.01745	Radians
Dynes	2.248×10^{-6}	Pound-force
Ergs.	1	Dyne-centimeters
Ergs.	7.376×10^{-8}	Foot pound-force
Ergs.	10^{-7}	Joules

COMMONLY USED CONVERSION FACTORS *(cont.)*

Multiply	By	To Obtain
Fathoms	6	Feet
Feet..................	30.48	Centimeters
Feet of air	3.608×10^{-5}	Atmospheres
Feet of air	0.0009	Feet of mercury
Feet of air	0.00122	Feet of water
Feet of air	0.00108	Inches of mercury
Feet of air	0.00053	Pound per square in.
Feet of mercury	30.48	Centimeters of mercury
Feet of mercury	13.6086	Feet of water
Feet of mercury	163.30	Inches of water
Feet of mercury	5.8938	Pound per square in.
Feet of water	0.0295	Atmospheres
Feet of water	2.2419	Centimeters of mercury
Feet of water	29888.9	Dynes/square cm
Feet of water	0.8826	Inches of mercury
Feet of water	304.78	Kg. per square meter
Feet of water	2988.888	Pascals
Feet of water	62.424	Pounds per square ft.
Feet of water	0.4335	Pounds per square in.
Foot pound-force	0.001286	British thermal units
Foot pound-force	5.050×10^{-7}	Horsepower-hours
Foot pound-force	1.356	Joules
Foot pound-force	0.1383	Kilogram-meters
Foot pound-force	3.766×10^{-7}	Kilowatt-hours
Gallons	3785.41	Cubic centimeters
Gallons	0.1337	Cubic feet
Gallons	231	Cubic inches
Gallons	0.003785	Cubic meters
Gallons	0.00495	Cubic yards
Gallons	3.7854	Liters
Gallons	128	Ounces
Gallons	8	Pints
Gallons	4	Quarts

COMMONLY USED CONVERSION FACTORS (cont.)

Multiply	By	To Obtain
Gallons per hour........	0.13368	Cubic feet per hour
Gallons per minute......	8.0208	Cubic feet per hour
Gallons per minute......	0.002228	Cubic feet per sec.
Horsepower	42.375	B.t.u. per min.
Horsepower	2,542.5	B.t.u. per hour
Horsepower	550	Foot pounds per sec.
Horsepower	33,000	Foot pounds per min.
Horsepower	1.014	Horsepower (metric)
Horsepower	10.686	Kg. calories per min.
Horsepower	0.7457	Kilowatts
Horsepower (boiler)	33,445.7	B.t.u. per hour
Horsepower-hours	2,546.1	British thermal units
Horsepower-hours	1.98×10^6	Foot pound-force
Horsepower-hours	2.737×10^5	Kilogram-meters
Horsepower-hours	0.7457	Kilowatt-hours
Inches................	2.540	Centimeters
Inches................	0.08333	Feet
Inches of mercury.......	0.03342	Atmospheres
Inches of mercury.......	1.133	Feet of water
Inches of mercury.......	3386.39	Pascals
Inches of mercury.......	70.526	Pounds per square ft.
Inches of mercury.......	0.4912	Pounds per square in.
Inches of water........	0.002458	Atmospheres
Inches of water........	2,490.8	Dynes per square cm
Inches of water........	0.07355	Inches of mercury
Inches of water........	25.398	Kg. per square meter
Inches of water........	0.5781	Ounces per square in.
Inches of water........	5.202	Pounds per square ft.
Inches of water........	0.03613	Pounds per square in.
Joules................	9.478×10^{-4}	British thermal units
Joules................	0.2388	Calories
Joules................	10^7	Ergs
Joules................	0.7376	Foot-pounds

COMMONLY USED CONVERSION FACTORS (cont.)

Multiply	By	To Obtain
Joules	2.778×10^{-7}	Kilowatt-hours
Joules	0.1020	Kilogram-meters
Joules	1	Watt-seconds
Kilograms	2.205	Pounds
Kilogram-calories	3.968	British thermal units
Kilogram meters	7.233	Foot pound-force
Kg per square meter	0.003281	Feet of water
Kg per square meter	0.2048	Pounds per square ft.
Kg per square meter	0.001422	Pounds per square in.
Kilometers	3280.84	Feet
Kilometers	0.6214	Miles
Kilopascals	0.009869	Atmospheres
Kilopascals	0.2952	Inches of mercury
Kilopascals	4.021	Inches of water
Kilopascals	0.010197	Kg. per square cm
Kilopascals	1,000	Pascals
Kilopascals	20.88542	Pounds per square ft.
Kilopascals	0.1450377	Pounds per square in.
Kilowatts	56.8725	B.t.u. per min.
Kilowatts	737.56	Foot pounds per sec.
Kilowatts	1.341	Horsepower
Kilowatts-hours	3409.5	British thermal units
Kilowatts-hours	2.655×10^{6}	Foot pound-force
Knots	1.151	Miles
Liters	1000	Cubic centimeters
Liters	0.03531	Cubic feet
Liters	61.0237	Cubic inches
Liters	0.001	Cubic meters
Liters	0.00131	Cubic yards
Liters	0.2642	Gallons
Log N_e or in N	0.4343	Log_{10} N
Log N	2.303	Log_e N or in N
Lumens per square ft.	1	Footcandles

COMMONLY USED CONVERSION FACTORS (cont.)

Multiply	By	To Obtain
Megohms	10^6	Ohms
Meters.	100	Centimeters
Meters.	0.54681	Fathoms
Meters.	3.281	Feet
Meters.	39.37	Inches
Meters.	0.001	Kilometers
Meters.	1,000	Millimeters
Meters.	1.0936	Yards
Meter-kilograms	7.233	Foot pounds
Microhms	10^{-6}	Ohms
Miles	5,280	Feet
Miles	1.609	Kilometers
Miner's inch	1.5	Cubic feet per min.
Minutes (angle)	0.016667	Degrees
Minutes (angle)	1.85×10^{-4}	Quadrants
Minutes (angle)	2.909×10^{-4}	Radians
Newtons	10^5	Dynes
Newtons	1.0	Joules per meter
Newtons	7.233	Poundals
Newtons	0.22481	Pound-force
Ohms	10^{-6}	Megohms
Ohms	10^6	Microhms
Ohms per mil foot.	0.1662	Microhms per cm. cube
Ohms per mil foot.	0.06524	Microhms per in. cube
Pascals	9.8692×10^{-6}	Atmospheres
Pascals	3.3455×10^{-4}	Foot of water
Pascals	0.000145	Foot pounds per sq. in.
Pascals	2.953×10^{-4}	Inches of mercury
Pascals	0.0040146	Inches of water
Pascals	0.101972	Kg per sq. meter
Poundals.	0.03108	Pound-force
Pound-force	32.174	Poundals
Pound feet	1.488	Kilograms per meter

COMMONLY USED CONVERSION FACTORS (cont.)

Multiply	By	To Obtain
Pounds of water	0.01602	Cubic feet
Pounds of water	0.1198	Gallons
Pounds per cubic foot . . .	16.02	Kg. per cubic meter
Pounds per cubic foot . . .	5.787×10^{-4}	Pounds per cubic in.
Pounds per cubic inch . . .	27.68	Grams per cubic cm.
Pounds per cubic inch . . .	2.768×10^4	Kg. per cubic meter
Pounds per cubic inch . . .	1,728	Pounds per cubic ft.
Pounds per square foot . .	0.01602	Feet of water
Pounds per square foot . .	4.8824	Kg. per square meter
Pounds per square foot . .	0.006944	Pounds per sq. in.
Pounds per square inch . .	2.3067	Feet of water
Pounds per square inch . .	2.036	Inches of mercury
Pounds per square inch . .	144	Pounds per sq. ft.
Radians.	57.296	Degrees
Square centimeters	1.973×10^5	Circular mils
Square Feet	2.296×10^{-5}	Acres
Square Feet	0.0929	Square meters
Square inches.	1.273×10^6	Circular mils
Square inches.	6.4516	Square centimeters
Square Kilometers	0.3861	Square miles
Square meters	10.764	Square feet
Square miles.	640	Acres
Square miles.	2.590	Square kilometers
Square Milimeters.	1,973	Circular mils
Square mils.	1.273	Circular mils
Tons (long)	2,240	Pounds
Tons (metric).	2,204.6	Pounds
Tons (short).	2,000	Pounds
Watts	0.0568	B.t.u. per minute
Watts	10^7	Ergs per sec.
Watts	44.2537	Foot pounds per min.
Watts	0.001341	Horsepower
Watts	14.34	Calories per min.

COMMONLY USED CONVERSION FACTORS *(cont.)*

Multiply	By	To Obtain
Watt-hours	3.4144	British thermal units
Watt-hours	2,655.22	Foot pounds
Watt-hours	0.001341	Horsepower-hours
Watt-hours	0.8604	Kilogram-calories
Watt-hours	367.098	Kilogram-meters
Webers	10^8	Maxwells

POUNDS PER SQUARE FOOT TO KILOPASCALS

Pounds per Square Foot	Kilopascals	Pounds per Square Foot	Kilopascals
1	.0479	8	.3832
2	.0958	9	.4311
3	.1437	10	.4788
4	.1916	25	1.1971
5	.2395	50	2.3940
6	.2874	75	3.5911
7	.3353	100	4.7880

POUNDS PER SQUARE INCH TO KILOPASCALS

Pounds per Square Inch	Kilopascals	Pounds per Square Inch	Kilopascals
1	6.895	8	55.160
2	13.790	9	62.055
3	20.685	10	68.950
4	27.580	25	172.375
5	34.475	50	344.750
6	41.370	75	517.125
7	48.265	100	689.500

DECIMAL EQUIVALENTS OF FRACTIONS

8ths	32nds	64ths	64ths
$\frac{1}{8}$ = .125	$\frac{1}{32}$ = .03125	$\frac{1}{64}$ = 0.15625	$\frac{33}{64}$ = .515625
$\frac{1}{4}$ = .250	$\frac{3}{32}$ = .09375	$\frac{3}{64}$ = .046875	$\frac{35}{64}$ = .546875
$\frac{3}{8}$ = .375	$\frac{5}{32}$ = .15625	$\frac{5}{64}$ = .078125	$\frac{37}{64}$ = .57812
$\frac{1}{2}$ = .500	$\frac{7}{32}$ = .21875	$\frac{7}{64}$ = .109375	$\frac{39}{64}$ = .609375
$\frac{5}{8}$ = .625	$\frac{9}{32}$ = .28125	$\frac{9}{64}$ = .140625	$\frac{41}{64}$ = .640625
$\frac{3}{4}$ = .750	$\frac{11}{32}$ = .34375	$\frac{11}{64}$ = .171875	$\frac{43}{64}$ = .671875
$\frac{7}{8}$ = .875	$\frac{13}{32}$ = .40625	$\frac{13}{64}$ = .203128	$\frac{45}{64}$ = .703125
16ths	$\frac{15}{32}$ = .46875	$\frac{15}{64}$ = .234375	$\frac{47}{64}$ = .734375
$\frac{1}{16}$ = .0625	$\frac{17}{32}$ = .53125	$\frac{17}{64}$ = .265625	$\frac{49}{64}$ = .765625
$\frac{3}{16}$ = .1875	$\frac{19}{32}$ = .59375	$\frac{19}{64}$ = .296875	$\frac{51}{64}$ = .796875
$\frac{5}{16}$ = .3125	$\frac{21}{32}$ = .65625	$\frac{21}{64}$ = .328125	$\frac{53}{64}$ = .828125
$\frac{7}{16}$ = .4375	$\frac{23}{32}$ = .71875	$\frac{23}{64}$ = .359375	$\frac{55}{64}$ = .859375
$\frac{9}{16}$ = .5625	$\frac{25}{32}$ = .78125	$\frac{25}{64}$ = .390625	$\frac{57}{64}$ = .890625
$\frac{11}{16}$ = .6875	$\frac{27}{32}$ = .84375	$\frac{27}{64}$ = .421875	$\frac{59}{64}$ = .921875
$\frac{13}{16}$ = .8125	$\frac{29}{32}$ = .90625	$\frac{29}{64}$ = .453125	$\frac{61}{64}$ = .953125
$\frac{15}{16}$ = .9375	$\frac{31}{32}$ = .96875	$\frac{31}{64}$ = .484375	$\frac{63}{64}$ = .984375

CONVERSION TABLE FOR TEMPERATURE – °F / °C

°F	°C	°F	°C	°F	°C	°F	°C	°F	°C
-459.4	-273	-22.0	-30	35.6	2	93.2	34	150.8	66
-418.0	-250	-18.4	-28	39.2	4	96.0	36	154.4	68
-328.0	-200	-14.8	-26	42.8	6	100.4	38	158.0	70
-238.0	-150	-11.2	-24	46.4	8	104.0	40	161.6	72
-193.0	-125	-7.6	-22	50.0	10	107.6	42	165.2	74
-148.0	-100	-4.0	-20	53.6	12	111.2	44	168.8	76
-130.0	-90	-0.4	-18	57.2	14	114.8	46	172.4	78
-112.0	-80	3.2	-16	60.8	16	118.4	48	176.0	80
-94.0	-70	6.8	-14	64.4	18	122.0	50	179.6	82
-76.0	-60	10.4	-12	68.0	20	125.6	52	183.2	84
-58.0	-50	14.0	-10	71.6	22	129.2	54	186.8	86
-40.0	-40	17.6	-8	75.2	24	132.8	56	190.4	88
-36.4	-38	21.2	-6	78.8	26	136.4	58	194.0	90
-32.8	-36	24.8	-4	82.4	28	140.0	60	197.6	92
-29.2	-34	28.4	-2	86.0	30	143.6	62	201.2	94
-25.6	-32	32.0	0	89.6	32	147.2	64	204.8	96

°F	°C	°F	°C	°F	°C	°F	°C	°F	°C
208.4	98	347.0	175	590	310	1004	540	6332	3500
212.0	100	356.0	180	608	320	1040	560	7232	4000
221.0	105	365.0	185	626	330	1076	580	4500	8132
230.0	110	374.0	190	644	340	1112	600	9032	5000
239.0	115	383.0	195	662	350	1202	650	9932	5500
248.0	120	392.0	200	680	360	1292	700	10832	6000
257.0	125	410	210	698	370	1382	750	11732	6500
266.0	130	428	220	716	380	1472	800	12632	7000
275.0	135	446	230	734	390	1562	850	13532	7500
284.0	140	464	240	752	400	1652	900	14432	8000
293.0	145	482	250	788	420	1742	950	15332	8500
302.0	150	500	260	824	440	1832	1000	16232	9000
311.0	155	518	270	860	460	2732	1500	17132	9500
320.0	160	536	280	896	480	3632	2000	18032	10000
329.0	165	554	290	932	500	4532	2500		
338.0	170	572	300	968	520	5432	3000		

1 degree F is 1/180 of the difference between the temperature of melting ice and boiling water.
1 degree C is 1/100 of the difference between the temperature of melting ice and boiling water.
Absolute Zero = -273.16°C = -459.69°F

TRIGONOMETRIC FORMULAS – RIGHT TRIANGLE

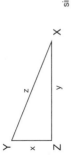

Angles = X, Y, Z
Distances = x, y, z
Area = $\dfrac{x\,y}{2}$

$\sin X = \dfrac{x}{z}$ $\cos X = \dfrac{y}{z}$

$\tan X = \dfrac{x}{y}$ $\cot X = \dfrac{y}{x}$

Pythagorean Theorem states
That $x^2 + y^2 = z^2$

Thus $x = \sqrt{z^2 - y^2}$

Thus $y = \sqrt{z^2 - x^2}$

Thus $z = \sqrt{x^2 + y^2}$

Given X and z, find Y, x, and y

$Y = 90° - X$, $x = z \sin X$, $y = z \cos X$

Given X and z, find Y, x, and z

$Y = 90° - X$, $x = y \tan X$, $z = \dfrac{y}{\cos X}$

Given X and x, find Y, y, and z

$Y = 90° - X$, $y = x \cot X$, $z = \dfrac{x}{\sin X}$

Given X and z, find X, Y, and y

$\sin X = \dfrac{x}{z} = \cos Y$, $y = \sqrt{(z^2 - x^2)} = z \sqrt{1 - \dfrac{x^2}{z^2}}$

Given x and y, find X, Y, and z

$\tan X = \dfrac{x}{y} = \cot Y$, $z = \sqrt{x^2 + y^2} = x \sqrt{1 + \dfrac{y^2}{x^2}}$

TRIGONOMETRIC FORMULAS – OBLIQUE TRIANGLES

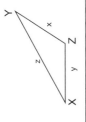

Given x, y, and z, Find X, Y, and Z

$$s = \frac{x + y + z}{2}, \quad \sin \frac{1}{2} X = \sqrt{\frac{(s-y)(s-z)}{yz}}$$

$$\sin \frac{1}{2} Y = \sqrt{\frac{(s-x)(s-z)}{xz}}, \quad C = 180° - (X + Y)$$

Given x, y, and z, find the Area

$$s = \frac{x + y + z}{2}, \quad \text{Area} = \sqrt{S(s-x)(s-y)(s-z)}$$

$$\text{Area} = \frac{yz \sin X}{2}, \quad \text{Area} = \frac{x^2 \sin Y \sin Z}{2 \sin X}$$

Given x, y, and Z, find X, Y, and z

$$X + Y = 180° - Z, \quad z = \frac{x \sin Z}{\sin X}, \quad \tan X = \frac{x \sin Z}{y - (x \cos Z)}$$

Given X, x, and y, Find Y, Z, and z

$$\sin Y = \frac{y \sin X}{x}, \quad Z = 180° - (X+Y), \quad z = \frac{x \sin Z}{\sin X}$$

Given X, Y, and x, Find y, Z, and z

$$y = \frac{x \sin Y}{\sin X}, \quad Z = 180° - (X+Y), \quad z = \frac{x \sin Z}{\sin X}$$

9-13

TRIGONOMETRIC FORMULAS – SHAPES

Equilateral Triangle	Annulus	Trapezium

Equilateral Triangle

X = Sides (Equal Lengths)

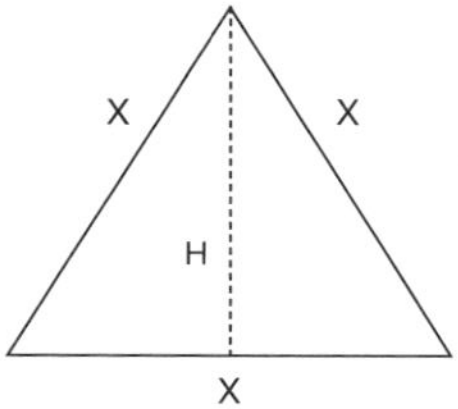

$Area = X^2 \sqrt{3/4} = .433\,X^2$

$Perimeter = 3\,X$

$H = X/2 \sqrt{3} = .866\,X$

Annulus

C_1 and R_1 = Inside Circle

C_2 and R_2 = Outside Circle

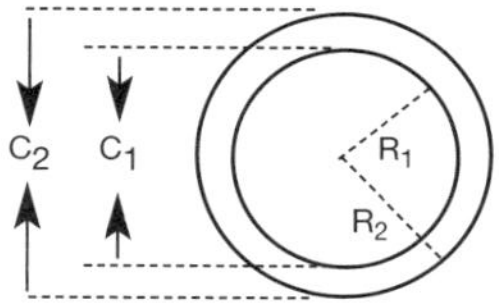

C = Circumference

R = Radius

$Area = \pi\,(R_1 + R_2)\,(R_2 - R_1)$

$Area = \left((C_2)^2 - (C_1)^2\right).7854$

Trapezium

Perimeter is the Sum of L, M, N, and O

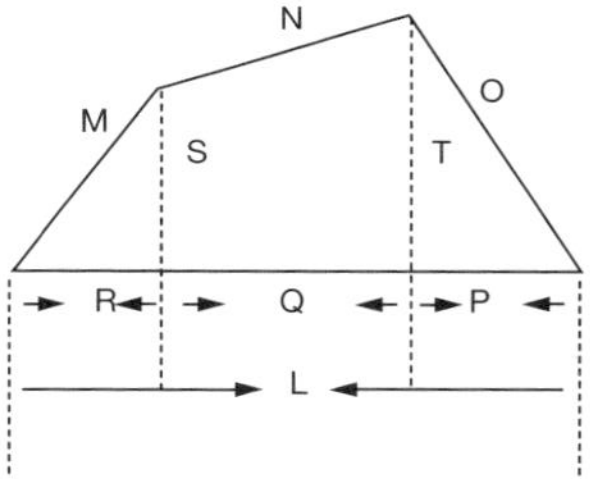

$Area = \dfrac{(S + T)\,Q + RS + PT}{2}$

9-14

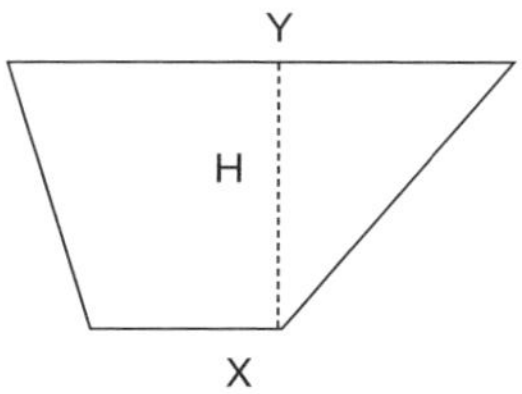

Quadrilateral

$Area =$
$$\frac{L_1 \cdot L_2 \cdot \sin \theta}{2}$$

Where θ = Degrees of Angle

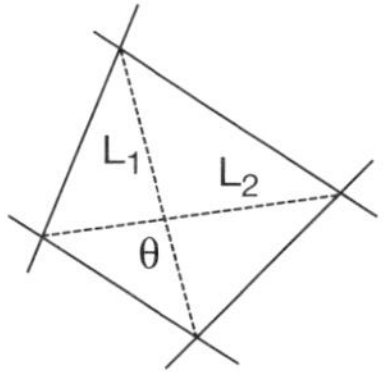

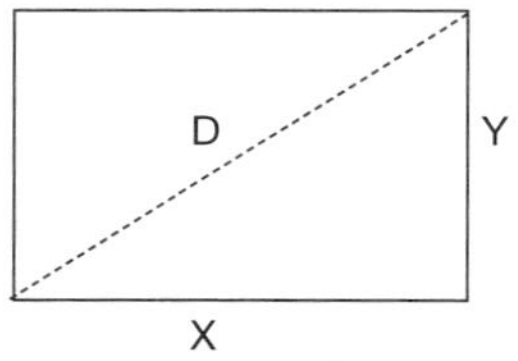

Parallelogram

Where θ = Degrees of Angle

$Area =$
$XH = XY \sin \theta$
$Perimeter =$
$2(X + Y)$

Trapezoid

$Perimeter =$
The Sum of the lengths of all four sides

$Area = \dfrac{(X + Y)}{2}$

Rectangle

$Area = XY$
Diagonal Line (D)
$= \sqrt{X^2 + Y^2}$
$Perimeter =$
$2(X + Y)$
If a square
then $X = Y$

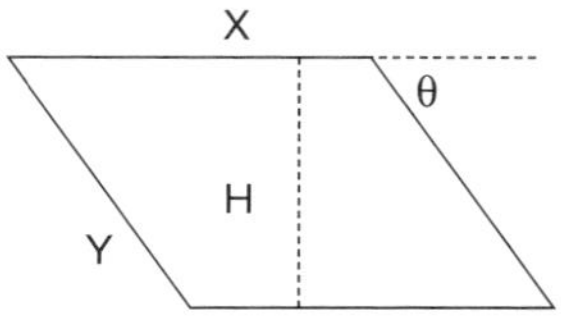

COMMON ENGINEERING UNITS AND THEIR RELATIONSHIP

Quantity	SI Metric Units/Symbols	Customary Units	Relationship of Units
Acceleration	meters per second squared (m/s²)	feet per second squared (ft./s²)	m/s² = ft./s² x 3.281
Area	square meter (m²) square millimeter (mm²)	square foot (ft.²) square inch (in.²)	m² = ft.² x 10.764 mm² = in.² x 0.00155
Density	kilograms per cubic meter (kg/m³) grams per cubic centimeter (g/cm³)	pounds per cubic foot (lb./ft.³) pounds per cubic inch (lb./in.³)	kg/m³ = lb./ft.³ x 16.02 g/cm³ = lb./in.² x 0.036
Work	Joule (J)	foot pound force (ft. lbf. or ft. lb.)	J = ft. lbf. x 1.356
Heat	Joule (J)	British thermal unit (BTU) Calorie (Cal)	J = BTU x 1.055 J = cal x 4.187
Energy	kilowatt (kW)	Horsepower (HP)	kW = HP x 0.7457

Quantity	SI Unit	Pound-force / kilogram-force	Conversion
Force	Newton (N) Newton (N)	Pound-force (lbf., lb.-f., or lb.) kilogram-force (kgf, kg · f., or kp)	$N = \text{lbf} \times 4.448$ $N = \dfrac{\text{kgf}}{9.807}$
Length	meter (m) millimeter (mm)	foot (ft.) inch (in.)	$m = \text{ft.} \times 3.281$ $mm = \dfrac{\text{in.}}{25.4}$
Mass	kilogram (kg) gram (g)	pound (lb.) ounce (oz.)	$kg = \text{lb.} \times 2.2$ $g = \dfrac{\text{oz.}}{28.35}$
Stress	Pascal = Newton per second (Pa = N/s)	pounds per square inch (lb./in.² or psi)	$Pa = \text{lb./in.}^2 \times 6{,}895$
Temperature	degree Celsius (°C)	degree Fahrenheit (F)	$^\circ C = \dfrac{^\circ F - 32}{1.8}$
Torque	Newton meter (N · m)	foot-pound (ft. lb.) inch-pound (in. lb.)	$N \cdot m = \text{ft. lbf.} \times 1.356$ $N \cdot m = \text{in. lbf.} \times 0.113$
Volume	cubic meter (m³) cubic centimeter (cm³)	cubic foot (ft.³) cubic inch (in.³)	$m^3 = \text{ft.}^3 \times 35.314$ $cm^3 = \dfrac{\text{in.}^3}{16.387}$

COMMONLY USED GEOMETRICAL RELATIONSHIPS

Diameter of a circle $\times$ 3.1416 = Circumference

Radius of a circle $\times$ 6.283185 = Circumference

Square of the radius of a circle $\times$ 3.1416 = Area

Square of the diameter of a circle $\times$ 0.7854 = Area

Square of the circumference of a circle $\times$ 0.07958 = Area

Half the circumference of a circle $\times$ half its diameter = Area

Circumference of a circle $\times$ 0.159155 = Radius

Square root of the area of a circle $\times$ 0.56419 = Radius

Circumference of a circle $\times$ 0.31831 = Diameter

Square root of the area of a circle $\times$ 1.12838 = Diameter

Diameter of a circle $\times$ 0.866 = Side of an inscribed equilateral triangle

Diameter of a circle $\times$ 0.7071 = Side of an inscribed square

Circumference of a circle $\times$ 0.225 = Side of an inscribed square

Circumference of a circle $\times$ 0.282 = Side of an equal square

Diameter of a circle $\times$ 0.8862 = Side of an equal square

Base of a triangle $\times$ one-half the altitude = Area

Multiplying both diameters and .7854 together = Area of an ellipse

Surface of a sphere $\times$ one-sixth of its diameter = Volume

Circumference of a sphere $\times$ its diameter = Surface

Square of the diameter of a sphere $\times$ 3.1416 = Surface

Square of the circumference of a sphere $\times$ 0.3183 = Surface

Cube of the diameter of a sphere $\times$ 0.5236 = Volume

Cube of the circumference of a sphere $\times$ 0.016887 = Volume

Radius of a sphere $\times$ 1.1547 = Side of an inscribed cube

Diameter of a sphere divided by $\sqrt{3}$ = Side of an inscribed cube

Area of its base $\times$ one-third of its altitude = Volume of a cone or pyramid whether round, square, or triangular

Area of one of its sides $\times$ 6 = Surface of the cube

Altitude of trapezoid $\times$ one-half the sum of its parallel sides = Area

CHAPTER 10
Symbols and Abbreviations

PIPE LEGEND COLOR CODING		
Classification	**Color of Band**	**Color of Letters**
Fire protection	Red	White
Dangerous	Yellow	Black
Safe	Green	Black
Protective	Blue	White

PIPE LEGEND SIZING IN INCHES		
Outside Diameter of Pipe or Covering	**Width of Color Band**	**Size of Legend Letters**
¾ to 1¼	8	½
1½ to 2	8	¾
2¼ to 6	12	1¼
8 to 10	24	2½
Over 10	32	3½

PIPING

Soil, waste or leader (above grade)	————————	Sanitary sewer	—— SS ——
Sanitary above grade	—— S ——	Storm drain	—— SD ——
		Storm above grade	—— ST ——
Soil, waste or leader (below grade)	— — — —	Storm below grade	— — ST — —
Sanitary below grade	- - - - S - - - -	Rain water leader	—— RWL ——
Vent	— — — — —	Sump pump discharge	—— SPD ——
Combination waste and vent	—— CWV ——	Service water	—— SWS ——
Waste line	—— W ——	Cold water	—— - —— - ——
Drain line	—— D ——	Soft cold water	—— SWC ——
Vent line	—— V ——	Drinking water supply	——DWS——
Indirect waste	—— IW ——	Drinking water return	——DWR——
Indirect vent	—— IV ——	Chilled drinking water supply	—— CDWS ——

Chilled drinking water return	—— CDWR ——
Hot water	—— · · —— · · ——
Hot water circulation	—— — — — —
Hot water return	— — · · — — · · —
Sanitizing hot water supply (180F)	—/— · · —/— · · —/
Sanitizing hot water return (180F)	—/— · · · —/— · · · —/
Industrialized hot water supply	—— IHW ——
Industrialized hot water return	—— IHR ——
Industrialized cold water	—— ICW ——
Industrial waste	—— INW ——
Tempered water, potable	—— T ——
Tempered water supply	—— TWS ——
Tempered water return	—— TWR ——
Acid water	—— ACID ——
Acid waste above grade	—— AW ——
Acid waste below grade	— — AW — —
Acid vent	— — AV — —
Chemical resistant waste	—— CRW ——
Fire line	— F —— F —
Wet standpipe	—— WSP ——
Dry standpipe	—— DSP ——
Combination standpipe	—— CSP ——
Main sprinkler supply	—— SPR ——

PIPING (cont.)

Branch and head sprinkler	——o——o——		Hydrogen	—— H ——
Gas – low pressure	— G — G —		Helium	—— HE ——
			Argon	—— AR ——
Gas – medium pressure	—— MG ——		Liquid petroleum gas	—— LPG ——
Gas – high pressure	—— HG ——		Pneumatic tubes tube runs	—— PN ——
Compressed air	—— CA ——		Cast iron	—— CI ——
Vacuum	—— VAC ——		Culvert pipe	—— CP ——
Vacuum cleaning	—— VC ——		Clay tile	—— CT ——
Oxygen	—— O ——		Ductile iron	—— DI ——
Liquid oxygen	—— LO ——		Reinforced concrete	—— RCP ——
Nitrogen	—— N ——		Drain – open tile or agricultural tile	= = = =
Liquid nitrogen	—— LN ——			
Nitrous oxide	—— NO ——		Low pressure steam	—— LPS ——

PIPING (cont.)

Medium pressure steam ——MPS——	Medium Temperature Hot Water Return ——MTWR——
High pressure steam ——HPS——	High Temperature Hot Water Return ——HTWR——
Low pressure return ——LPR——	Boiler blow down ——BBD——
Medium pressure return ——MPR——	Feedwater pump discharge ——FPD——
High pressure return ——HPR——	Hot water heating supply ——HWS——
Low Temperature Hot Water Supply ——HWS——	Hot water heating return ——HWR——
Medium Temperature Hot Water Supply ——MTWS——	Fuel oil suction ——FOS——
	Fuel oil return ——FOR——
High Temperature Hot Water Supply ——HTWS——	Fuel oil tank vent ——FOV——
	Existing Piping ——(NAME) E——
Low Temperature Hot Water Return ——HWR——	Pipe to be Removed ✕✕ (NAME) ✕✕

VALVES		
Type	**Screwed**	**Bell & Spigot**
Gate Valve		
Globe Valve		
Check Valve		
Angle Check Valve		
Stop Cock		
Relief or Safety Valve		

VALVES (*cont.*)

Angle Globe Valve		Solenoid Valve	
Angle Gate Valve		Diaphragm Operated Valve	
Quick Opening Valve		Reducing Valve (Self Actuated)	
Float Opening Valve		Reducing Valve (External Pilot Connection)	
Post Indicator Gate Valve		Lock shield	
Plug Valve		2-Way automatic control	
Plug Cock		3-Way automatic control	
Butterfly Valve		Gas cock	
Pressure Reducing Valve		Shock absorber	
Hose Gate		OS & Y gate	
Three-Way Valve		Strainer	
Motor Operated Valve		Temperature and pressure relief valve	
Motor Operated Gate Valve			

FITTINGS

Fitting connections are often specified with the fitting symbol. For example, an elbow could have the following types of connections:

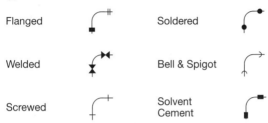

Flanged	Soldered
Welded	Bell & Spigot
Screwed	Solvent Cement

Fittings are shown with screwed connections unless specified.

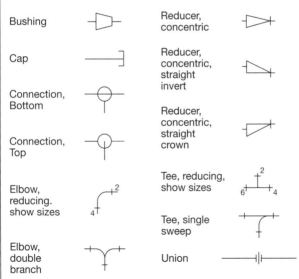

Bushing	Reducer, concentric
Cap	Reducer, concentric, straight invert
Connection, Bottom	Reducer, concentric, straight crown
Connection, Top	Tee, reducing, show sizes
Elbow, reducing. show sizes	Tee, single sweep
Elbow, double branch	Union

10-8

FITTINGS (cont.)		
Type	Screwed	Bell & Spigot
Joint, coupling		
Elbow–90°		
Elbow–45°		
Elbow–Turned Up		
Elbow–Turned Down		
Elbow–Long Radius		

FITTINGS (cont.)		
Type	**Screwed**	**Bell & Spigot**
Side Outlet Elbow – Outlet Down		
Side Outlet Elbow – Outlet Up		
Base Elbow		
Double Branch Elbow		
Single Sweep Tee		
Double Sweep Tee		

FITTINGS (cont.)		
Type	**Screwed**	**Bell & Spigot**
Reducing Elbow, Show Sizes		
Tee		
Tee – Outlet Up		
Tee – Outlet Down		
Side Outlet Tee – Outlet Up		
Side Outlet Tee – Outlet Down		

FITTINGS (cont.)		
Type	**Screwed**	**Bell & Spigot**
Cross		
Reducer		
Eccentric Reducer		
Lateral		
Expansion Joint Flanged		

SPECIALTIES AND MISCELLANEOUS

Alignment guide, pipe		Flanged joint		
Anchor, intermediate	PA	Flexible connection		
Anchor, main	PA	Flexible connector		
Ball joint		Flow direction		
Concentric reducer		Flowmeter, orifice	OFM-1	
Eccentric reducer		Flowmeter, venturi	VFM-1	
Elbow looking up		Flow switch	FS	
Elbow looking down		Funnel drain, open		
Expansion joint		Hanger, rod	H	
Expansion loop		Hanger, spring	H	

10-13

SPECIALTIES AND MISCELLANEOUS (cont.)

Pitch of pipe,
Rise (R)
Drop (D)

Pressure
gauge

Pressure
switch

Pressure
switch, dual
(high low)

Pump,
indicate use

Spool piece,
flanged

Strainer

Strainer,
blow off

Strainer,
duplex

Tank,
indicate use

Thermometer

Thermostat

Thermostat,
self-contained

Thermostat,
remote bulb

Thermostatic
trap

Float and
thermostatic
trap

House trap

'P' trap

Traps, steam
indicate type

'Y'

FIXTURES

Baths

Corner

Recessed

Roll Rim

Angle

Whirlpool

Institutional or island

SB

Sitz Bath

FB

Foot Bath

Showers

×

Stall

Corner Stall

(Plan)

(Elev.)

Shower head

(Plan)

(Elev)

Overhead gang
Shower heads

FIXTURES (cont.)

Water Closets

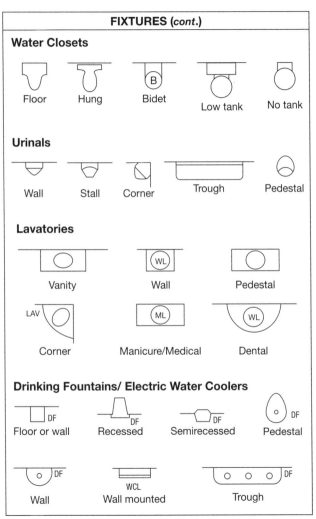

Floor   Hung   Bidet   Low tank   No tank

Urinals

Wall   Stall   Corner   Trough   Pedestal

Lavatories

Vanity   Wall   Pedestal

Corner   Manicure/Medical   Dental

Drinking Fountains/ Electric Water Coolers

Floor or wall   Recessed   Semirecessed   Pedestal

Wall   Wall mounted   Trough

FIXTURES (cont.)

Sinks/Dishwashers

Single basin

Right drainboard

Twin basin

Left drainboard

Plain kitchen sink

Double drainboard

Sink/Dishwasher
Combination

Dishwasher

Wash sink

Wash sink wall type

Service Sinks

Wall

Floor

Wash Fountains

Circular

Semicircular

FIXTURES (cont.)

Laundry Trays

Single	Double	Combination sink and laundry tray
LT	L T	S T

Hot Water

Heater	Tank
HW	HWT

Separators

Gas	Oil
G	O

Cleanouts

Cleanout	Cleanout	Floor	Wall
	C O		WCO

Drains

Drain	Floor drain	Floor drain with backup valve
D	FD	

Garage drain	Roof drain	Roof sump

FIXTURES (cont.)

Miscellaneous

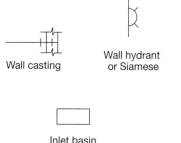

Wall casting

Wall hydrant
or Siamese

Manhole
(identify by number)

Inlet basin
(identify by number)

Catch basin
(identify by number)

Meter

Hose rack

Hose bib

Gas outlet

Vacuum outlet

FIRE PROTECTION

Piping

Fire protection water supply —— F ——	Pendant heads
Wet standpipe —— WSP ——	Flush mounted heads
Dry standpipe —— DSP ——	Sidewall heads
Combination standpipe —— CSP ——	Fire hydrant
Automatic fire sprinkler —— SP ——	
Drain —— D ——	Wall fire department connection
Riser and branch (give size) ⊗—○—○ 4 in.	Sidewalk fire department connection
Pipe hanger	
Control valve	Fire hose rack FHR
Alarm check valve	Surface mounted fire hose cabinet FHC
Dry pipe valve	
Upright fire sprinkler heads ○—○	Recessed fire hose cabinet FHC

FIRE PROTECTION (cont.)

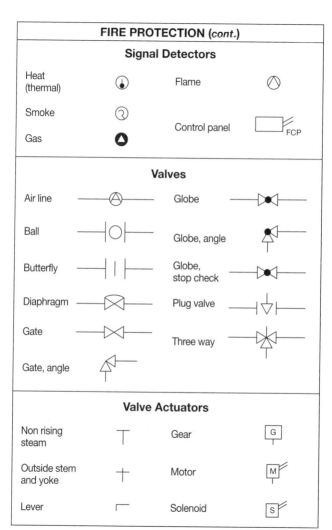

Signal Detectors

Heat (thermal)

Flame

Smoke

Control panel   FCP

Gas

Valves

Air line

Globe

Ball

Globe, angle

Butterfly

Globe, stop check

Diaphragm

Plug valve

Gate

Three way

Gate, angle

Valve Actuators

Non rising steam

Gear   G

Outside stem and yoke

Motor   M

Lever

Solenoid   S

Special Duty Valves

Check, swing gate			Pressure reducing, external pressure	
Check, spring			Pressure reducing, differential pressure	
Control, electric-pneumatic			Quick opening	
Control, pneumatic-electric			Quick closing, fusible link	
Hose end drain			Relief (R) or Safety (S)	
Lock shield			Solenoid	
Needle			Square head cock	
Pressure reducing, self-contained				

UNDERSTANDING PLAN SYMBOLS

Symbol	Definition
(XX)	Sheet note (number), applies only on the sheet it appears on
[XX]	Coordination point between floor plans and diagrams (number)
⟨X⟩	Demolition note (number)
(X / X)	Direction of view Section number Drawing on which section or detail is shown
(X / X)	Detail or section number Drawing from which section or detail is taken

UNDERSTANDING PLAN SYMBOLS (cont.)

Symbol	Definition

Equipment symbol (see schedule)

Equipment designation

System number if applicable

Equipment reference number

Equipment symbol (see schedule)

Equipment designation

Equipment reference number

Existing to remain

Existing to be removed

New work

Limits of work
(trade contracts, etc.)

ARCHITECTURAL SYMBOLS

Symbol	Definition

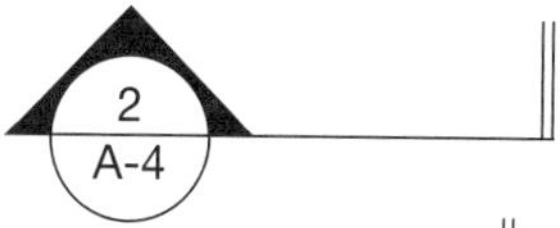

Wall section No. 2 can be seen on drawing No. A-4

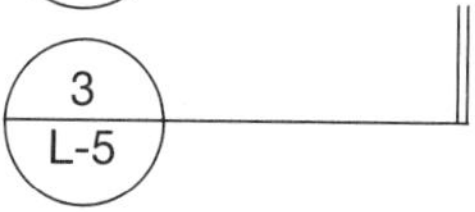

(symbol: filled triangle over circle containing "2" above "A-4")

Detail section No. 3 can be seen on drawing No. A-5.

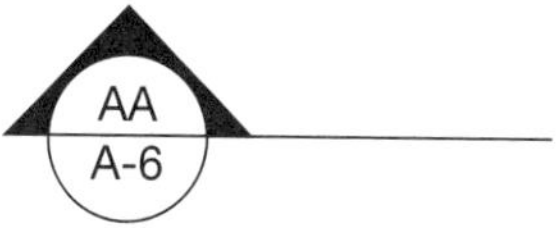

(symbol: circle containing "3" above "L-5")

Building section A-A can be seen on drawing No. A-6.

(symbol: filled triangle over circle containing "AA" above "A-6")

Main object line

Hidden or invisible line

Indicates center line

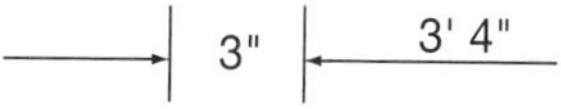

3"   3' 4"   Dimension lines

Extension lines

Symbol indicates center line

Indicates wall suface

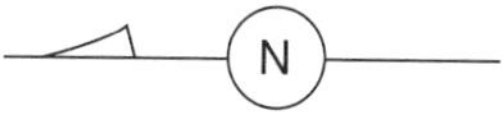

Indicates north direction

DRAWING CONVENTIONS AND SYMBOLS

Graphic Symbols

The symbols shown are those that seem to be the most common and acceptable, judged by the frequency of use by the architectural offices surveyed. This list can and should be expanded by each office to include symbols generally used by it, but not indicated here. Adoption of these symbols as standard practice is desirable to improve communication in the industry.

 Stair direction symbol

 North point
to be placed on each
floor plan, generally in
lower right hand corner
of drawings

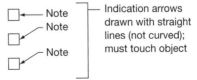 Indication arrows
drawn with straight
lines (not curved);
must touch object

Dash and dot
Center lines, projections, existing elevations lines

Dash and double dot line
Property lines, boundary lines

Dotted line
Hidden, future or existing construction to be removed

Break line
To break off parts of drawing

Linework

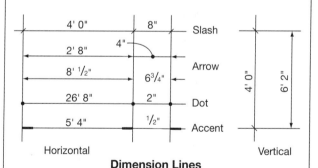

Horizontal		
4' 0"	8"	Slash
2' 8"	4"	Arrow
8' 1/2"	6 3/4"	
26' 8"	2"	Dot
5' 4"	1/2"	Accent

4' 0"   6' 2"

Vertical

Dimension Lines

ABBREVIATIONS

A	compressed air line	**ASTM**	American Society for Testing Material
ABC	above ceiling	**ATM**	atmosphere
ABS	absolute or acrylonitrile butadiene styrene	**ATC**	at ceiling or automatic temperature control
AC	air chamber		
AFF	above finish floor	**AV**	acid resistant vent
AGA	American Gas Association	**AW**	acid resistant waste
ANMC	American National Metric Council	**AWWA**	American Water Works Association
ANSI	American National Standards Institute	**B & S**	bell-and-spigot (cast iron) pipe
AP	access panel	**B**	bidet
APOTH	apothecary	**BOCA**	Building Officials Conference of America
ASHRAE	American Society of Heating, Refrigerating and Air Conditioning Engineers	**BP**	back pressure
		BS	bar sink
		BT	bathtub
ASME	American Society of Mechanical Engineers	**BTU**	british thermal unit
		BV	butterfly valve
ASPE	American Society of Plumbing Engineers	**°C**	degrees Celsius
		C	condensate line
ASSE	American Society of Sanitary Engineers or American Society of Safety Engineers	**C to C**	center to center
		CAL	calorie
		CB	catch basin (identify by number)
		CD	condensate drain

ABBREVIATIONS (cont.)

CFH	cubic feet per hour	DS	down spout
CI	cast iron	DW	dishwasher
CIRC	circular	DWG	drawing
CISPI	Cast Iron Soil Pipe Institute	DWV	drainage, waste., and vent system
CISP	cast iron soil pipe	E to C	end to center
CM	centimeter	EJ	expansion joint
CM²	square centimeter	ET	expansion tank
CM³	cubic centimeter	EWC	electric water cooler
CO	cleanout	°F	degrees Fahrenheit
CPVC	chlorinated polyvinyl chloride	F	Fahrenheit or fire line
CRW	chemical resistant waste	FB	foot bath
CU	cubic	FCO	floor cleanout
CU FT.	cubic feet	FD	floor drain
CU IN.	cubic inches	FDC	fire department connection
CV or CKV	check valve	FEC	fire extinguisher cabinet
CW	cold water	FF	finish floor
CWR	cold water riser	FG	finish grade
CWM	clothes washing machine	FHC	fire hose cabinet
D	drain line	FL	flow meter
DB	decibel	FM	fire line
DF	drinking fountain	FOV	flush out valve
DM	decimeter	FP	fire plug
DM²	square decimeter	FPM	(feet) per minute
DM³	cubic decimeter	FS	federal specifications
DR	drain		

ABBREVIATIONS (cont.)

FSP	fire standpipe	**ID**	inside diameter
FT	feet	**IN.**	inch
FTG	fitting	**INHg**	inches of mercury
FU	fixture unit	**INT**	international
FV	flush valve	**IPS**	iron pipe size
G	gram or gas line	**IV**	indirect vent
GAL	gallons	**IW**	indirect waste
GALV	galvanized	**J**	joule
GC	general contractor	**K**	Kelvin
GL.V	globe valve	**KG**	kilogram
GPD	gallons per day	**KM**	kilometer
GPM	gallons per minute	**KM²**	square kilometer
GS	glass sink	**KPA**	kilopascal
GV	gate valve	**KS**	kitchen sink
GWH	gas water heater	**KW**	kilowatt
H₂O	water	**L**	length or liter
HB	hose bibb	**L or LAV**	lavatory
Hd/HD	head	**LB.**	pound
Hg	mercury	**LBF**	pound-force
HG	hose gate	**LIQ**	liquid
HR	hour	**LOG**	logarithm
HW	hot water	**LT**	laundry tray
HWH	hot-water heater	**M**	meter
HWR	hot water return or hot water riser	**M²**	square meter
		M TYPE	lightest type of rigid copper pipe
HWT	hot water tank		
IB	inlet basin (identify by number)	**MAX**	maximum

ABBREVIATIONS (cont.)

MCA	Mechanical Contractors Association	NSF	National Sanitation Foundation Testing Laboratory
MFR	manufacturer		
MG	milligram	O	oxygen
MGD	million gallons per day	OC	on center
		OD	outside diameter
MI	malleable iron	OED	open end drain
MIN	minimum or minute	OF	overflow
MH	manhole (identify by number)	OZ.	ounce
		PA	pascal
ML	milliliter	PC	plumbing contractor
MM	millimeter		
MM³	cubic millimeter	PD	planter drain
MPT	male pipe thread	PG	pressure gauge
MS	mild steel	PP	pool piping
NAPHCC	National Association of Plumbing Heating and Cooling Contractors	PRV	pressure reducing valve
		PSI	pounds per square inch
NBFU	National Board of Fire Underwriters	PSIG	pounds per square inch – gauge
NBS	National Bureau of Standards	PV	plug valve
NFPA	National Fire Protection Association	RAD	radius
		RD	roof drain
		RED	reducer
N-M	newton-meter	RL	roofleader
NPS	nominal pipe size	RV	relief valve
		RWL	rain water leader

ABBREVIATIONS (cont.)

SAN	sanitary	**TD**	temperature differential
S	soil line		
SA	shock absorber	**U or UR**	urinal (wall hung or stall)
SB	sitz bath		
SD	storm drainage	**UF**	under floor
SEC	second	**UH**	unit heater
SH	shower	**V**	vent line
SH HD	shower head	**VAN**	vanity
SI	International System of Units (metric)	**VCI**	vacuum cleaning inlet
		VCL	vacuum cleaning line
SHWR	shower	**VR**	vent riser
SPEC	specification	**VS**	vent stack
SQ.	square	**VTR**	vent through roof
SQ. FT.	square feet	**W**	waste or waste line
SP	swimming pool		
SP-1	sump pump and number	**WC**	water closet (floor or hung)
SPD	sump pump discharge	**WCL**	water cooler (wall mounted)
SS	service sink or sanitary sewer	**WCO**	wall cleanout
		WF	circular wash fountain
STD	standard		
SV	service	**WH**	water heater
SW	service weight	**WH-1**	water heater and number
SWS	service water		
S & W	soil and waste	**WM**	washing machine
T	tempered water, potable or temperature	**XH**	extra heavy

CHAPTER 11
Glossary

GLOSSARY

A

Absolute Pressure: the total pressure measured from absolute vacuum; the sum of gauge pressure and atmospheric pressure corresponding to the barometer expressed in pounds per square inch

Absolute Zero: a point of total absence of heat, equivalent to minus 273.18°C

Accessible: having access which first may require the removal of an access panel, door, etc.

Accumulator: a container in which fluid or gas is stored under pressure as a source of power

Acid Vent: a pipe venting an acid waste system

Acid Waste: a pipe which conveys liquid waste matter containing a pH of less than 7.0

Acme Thread: a screw thread, used extensively for feed screws

Acrylonitrile-Butadiene-Styrene: a thermoplastic compound from which fittings, pipe, and tubing are made

Active Sludge: sewage sediment, rich in destructive bacteria, that can be used to break down fresh sewage

Adapter Fitting: fittings designed to mate or fit two pipes or fittings which are different in design when connecting the two together would otherwise not be possible; a fitting that serves to connect two different tubes or pipes to each other, such as copper to iron

Administrative Authority: the individual official, board, department, or agency authorized by a state, county, city, or other political subdivision to administer and enforce the provisions of the plumbing code

Aeration: an artificial method in which water and air are brought into direct contact with each other; also used to furnish oxygen to water that is oxygen deficient by spraying the liquid in the air, bubbling air through the liquid or by agitation of the liquid

Aerobic: bacteria living or active only in the presence of free oxygen

Air Break: a physical separation in which a drain indirectly discharges into a fixture, receptacle or interceptor at a point below the rim of the receptacle to prevent backflow or back-siphonage

Air Chamber: a continuation of the water piping beyond the branch to fixtures finished with a cap designed to eliminate shock or vibration of the piping when a valve is closed suddenly

Air, Compressed: air at any pressure greater than atmospheric pressure

Air, Free: air which is not contained and which is subject only to atmospheric conditions

Air Gap: the unobstructed vertical distance through the free atmosphere between the lowest opening from any pipe or faucet conveying water or waste to a tank, plumbing fixture receptor or other device and the flood level rim of the receptacle normally twice the diameter of the inlet

Air, Standard: air having a temperature of 70°F, (21.1°C), a standard density of 0.0075 lb./ft. (0.11 kg/m), and under pressure of 14.70 psia. (101.4 kPa); the gas industry standard is 60°F (15.6°C)

Air Test: a test that is applied to the plumbing system upon its completion, but before the walls are closed

Alarm Check Valve: a check valve equipped with a signaling device which will annunciate a remote alarm that a sprinkler head or heads are discharging

Alloy: a substance composed of two or more metals or a metal and nonmetal

Alloy Pipe: a steel pipe with one or more elements other than carbon which give it greater resistance to corrosion and more strength than carbon steel pipe

Ambient Temperature: the prevailing temperature of the area surrounding an object

American Standard Pipe Thread: a type of screw thread commonly used on pipe and fittings

Anaerobic: bacteria living or active in the absence of free origin

Anchor: a device used to secure pipes to a building or structure

Angle of Bend: in a pipe, the angle between radial lines from the beginning and end of the bend to the center

Angle Valve: a device in which the inlet and outlet are at right angles

Approved: accepted or acceptable under an applicable specification or standard stated or cited for the proposed use by the administrative authority

Approved Testing Agency: an organization established for purposes of testing to approved standards and acceptable by the administrative authority

Area Drain: a receptacle designed to collect surface or rain water from a determined open area

Arterial Vent: a vent serving a building drain and a public sewer

Aspirator: a fitting or device supplied with water or other fluid under positive pressure which passes through an integral orifice or constriction causing a vacuum

Atmospheric Vacuum Breaker: a mechanical device consisting of a check valve opening to the atmosphere when the pressure in the piping drops to atmospheric

Authority Having Jurisdiction: the organization, office or individual responsible for approving equipment, installation or procedure

B

Backflow: the flow of water or other liquids, mixtures or substances into the distributing pipes of a potable water supply from any source other than its intended source

Backflow Connection: a condition in any arrangement where backflow may occur

Backflow Preventer: a device or means to prevent backflow into the potable water system

Backing Ring: a metal strip used to prevent melted metal from entering a pipe when making a butt-welded joint

Back-Siphonage: the flowing back of used or contaminated water from a fixture or vessel into a water supply pipe due to negative pressure in such pipe

Back Up: a condition where the waste water may flow back into another fixture or compartment but not into the potable water system

Backwater Valve: a device which permits drainage in one direction but has a check valve that closes against back pressure; sometimes used together with gate valves for sewage applications

Baffle Plate: a tray or partition placed in process equipment to change the direction of flow

Ball Check Valves: a device used to stop flow in one direction while allowing flow in an opposite direction

Ball Valve: a spherical shaped gate valve providing a very tight shut-off

Base: the lowest portion or lowest point of a stack of vertical pipe

Battery of Fixtures: two or more similar adjacent fixtures which discharge into a common horizontal waste or soil branch

Bell: that portion of a pipe which is sufficiently enlarged to receive the end of another pipe of the same diameter for making a joint

Bell and Spigot Joint: commonly used joint in cast iron soil pipe; each piece is made with an enlarged diameter or bell at one end into which the plain or spigot end of another piece is inserted and the joint is sealed by cement, oakum, lead or rubber caulked into the bell around the spigot

Black Pipe: non-galvanized steel pipe

Blank Flange: a soil plate flange used to seal off the flow in a pipe

Blind Flange: a flange used to seal off the end of a pipe

Boiler Blow-Off: an outlet on a boiler to permit discharge of sediment

Boiler Blow-Off Tank: a vessel designed to receive the discharge from a boiler blow-off outlet

Bonnet: that part of a valve which connects the valve actuator to the valve body

Branch: any part of the piping system other than a main, riser or stack

Branch Interval: a length of soil or waste stack corresponding to a story height within which the horizontal branches are connected to the stack

Branch Tee: a tee having one side branch

Branch Vent: a vent connecting one or more individual vents with a vent stack or stack vent

Brazed: joined by hard solder

Brazing Ends: the ends of a valve or fitting which are prepared for silver brazing

Bronze Trim or Bronze Mounted: indicates that certain internal parts of the valves known as trim materials (stem, disc, seat rings, etc.) are made of copper alloy

GLOSSARY *(cont.)*

BTU: abbreviation for British Thermal Unit; the amount of heat required to raise the temperature of one pound (0.45 kg) of water one degree fahrenheit (0.565°C)

Building Drain: that part of the lowest piping of the drainage system which receives the discharge from soil, waste, etc. inside the walls of the building and conveys it to the building sewer

Building Sewer: that part of the horizontal piping of a drainage system which extends from the end of the building drain and conveys it to a public sewer, private sewer or individual sewage-disposal system

Building Sewer, Combined: a building sewer which conveys both sewage and storm water

Building Sewer, Sanitary: a building sewer which conveys sewage only

Building Sewer, Storm: a building sewer which conveys storm water or other drainage but no sewage

Building Subdrain: that portion of a drainage system which cannot drain by gravity in the building sewer

Building Trap: a fitting or assembly of fittings installed in the building drain to prevent circulation of air between the drainage of the building and the building sewer; usually installed as a running trap

Bubble Tight: the condition of a valve seat that prohibits the leakage of visible bubbles when closed

Bull Head Tee: a branch of the tee is larger than the run

Burst Pressure: the pressure which can be slowly applied to the valve at room temperature for 30 seconds without causing rupture

Bushing: a pipe fitting for connecting a pipe with a female or larger size fitting; it has a hollow plug with internal and external threads

Butt-Weld Joint: a welded pipe joint made with the ends of two pipes butting each other

Butt-Weld Pipe: pipe welded along a seam butted edge to edge and not scarfed or lapped

Butterfly Valve: a device deriving its name from the wing-like action of the disk which operates at right angles to the flow

By-Pass: an auxiliary loop in a pipeline intended for diverting flow around a valve or other devices

By-Pass Valve: a valve used to divert the flow past the part of the system through which it normally passes

C

Capacity: the maximum or minimum flows possible under given conditions of media, temperature, pressure, velocity, etc.

Capillary: the action by which the surface of a liquid, where it is in contact with a solid, is elevated or depressed depending upon the relative attraction of the molecules of the liquid for each other and for those of the solid

Carbon Steel Pipe: steel pipe which owes its properties mostly to the carbon it contains

Cathodic Protection: the control of the electrolytic corrosion of an underground or underwater metallic structure by the application of an electric current in such a way that the structure is made to act as the cathode instead of the anode of an electrolytic cell

Cavitation: a localized gaseous condition that is found within a liquid stream

Cement Joint: the union of two fittings by insertion of material

Cesspool: a lined excavation in the ground which receives the discharge of a drainage system so as to retain the organic matter and solids but permitting the liquids to seep through the bottom and sides

Chainwheel Operated Valve: a device which is operated by a chain driven wheel which opens and closes valve seats

Chase: a recess in a wall in which pipes can be run

Check Valve: a device designed to allow a fluid to pass through in one direction only

Chemical Waste System: piping which conveys corrosive or harmful wastes to the drainage system

Circuit: the directed route taken by a flow from one point to another

Circuit Vent: a branch vent that serves two or more traps and extends from in front of the last fixture connection of a horizontal branch to the vent stack

Clamp Gate Valve: a gate valve whose body and bonnet are held together by a U-bolt clamp

Cleanout: a plug or cover joined to an opening in a pipe which can be removed for the purpose of cleaning or examining the pipe

Clear Water Waste: cooling water and condensate drainage from refrigeration and air conditioning equipment, cooled condensate from steam heating systems; cooled boiler blowdown water, waste water drainage from equipment rooms and other areas where water is used without an appreciable addition of oil, gasoline, solvent, acid, etc., and treated effluent in which impurities have been reduced below a minimum concentration considered harmful

Close Nipple: a nipple with a length twice the length of a standard pipe thread

Cock: a form of valve having a hole in a tapered plug which is rotated to provide passageway for fluid

Coefficient of Expansion: the increase in unit length, area, or volume for 1 degree rise in temperature

Combined Waste and Vent System: a specially designed system of waste piping, embodying the horizontal wet venting of one or more floor sinks or floor drains by means of a common waste and vent pipe, adequately sized to provide free movement of air above the flow line of the drain

Common Vent: a vent which connects at the junction of two fixture drains and serves as a vent for both fixtures; also known as a dual vent

Companion Flange: a pipe flange to connect with another flange or with a flanged valve or fitting; it is attached to the pipe by threads, welding or other methods and differs from a flange which is an integral part of a pipe or fitting

Compression Joint: a multi-piece joint with cup shaped threaded nuts which, when tightened, compress tapered sleeves so that they form a tight joint in the periphery of the tubing they connect

Compressor: a mechanical device for increasing the pressure of air or gas

Condensate: water which has liquified from steam

Conductor: the piping from the roof to the building storm drain, combined sewer or other approved means of disposal and located inside of the building

Confluent Vent: a vent serving more than one fixture vent or stack vent

Continuous Vent: a vent that is a continuation of the drain to which it connects

GLOSSARY (cont.)

Continuous Waste: a continuous drain from two or three fixtures connected to a single trap

Control: a device used to regulate the function of a component or system

Corporation Cock: a stopcock screwed into the street water main to supply the house service connection

Coupling: a pipe fitting with female threads used to connect two pipes in a straight line

Critical Level: the point on a backflow prevention device or vacuum breaker conforming to approved standards usually stamped or marked CL or C/L on the device by the manufacturer which determines the minimum elevation above the flood level rim of the fixture or receptacle served at which the device may be installed; when a backflow prevention device does not bear critical level marking, the bottom of the vacuum breaker, combination valve or the bottom of any such approved device shall constitute the critical level

Cross: a pipe fitting with four branches in pairs, each pair on one axis, and the axis at right angles

Cross-Over: a pipe fitting with a double offset or shaped like the letter "U" with ends turned out; used to pass the flow of one pipe past another when the pipes are in the same plane

Cross Valve: a valve fitted on a transverse pipe so as to open communication between two parallel pipes

Cross Connection: any physical connection or arrangement between two otherwise separated piping systems, one of which contains potable water and the other water or other substance of unknown or questionable safety, whereby flow may occur from one system to the other, the direction of flow depending on the pressure differential between the two systems

Crown: the top of a trap

Crown Vent: a vent pipe connected at the uppermost point in the crown of a trap

Cup Weld: a pipe weld where one pipe is expanded on the end to allow the entrance of the end of the other pipe; the weld is then circumferential at the end of the expanded pipe

Curb Box: a device at the curb that contains a valve used to shut off a supply line, usually gas or water

D

Dampen: to check or reduce

Dead End: a branch leading from a soil, waste or vent pipe, building drain or building sewer which is terminated at a developed distance of 2 feet or more by means of a plug or other closed fitting

Department Having Jurisdiction: the administrative authority and includes any other law enforcement agency affected by any provision of the plumbing code

Developed Length: the length along the center line of the pipe and fittings

Dewpoint: the temperature of a gas or liquid at which condensation or evaporation occurs

Diaphragm: a flexible disk used to separate the control medium from the controlled medium and which actuates the valve stem

Diaphragm Control Valve: a control valve having a spring diaphragm actuator

Dielectric Fitting: a fitting having insulating parts or material that prohibits flow of electric current

Differential: the variance between two target values, one of which is the high value, the other being the low value

Digestion: the portion of the sewage treatment process where biochemical decomposition of organic matter takes place resulting in the formation of simple organic and mineral substances

Disk: that part of a valve which actually closes off the flow

Displacement: the volume or weight of a fluid displaced by a floating body

Domestic Sewage: the liquid and water borne wastes that are free from industrial wastes and permit satisfactory disposal without special treatment into the public sewer or a private sewage disposal system

Dosing Tank: a watertight tank in a septic system placed between the septic tank and the distribution box that is equipped with a pump or automatic siphon designed to discharge sewage intermittently to a disposal field

Double Disk: a two-piece disk used in a gate valve

Double Extra-Strong Pipe: a schedule of steel or wrought iron pipe weights in common use

Double Offset: two changes of direction installed in succession or series in continuous pipe

Double Ported Valve: a valve having two parts to overcome line pressure imbalance

Double Sweep Tee: a tee made with long radius curves between body and branch

Double Wedge: a device used in gate valves similar to double disk where the split wedges seal independently

Down: refers to piping running through the floor to a lower level

Downspout: the rainleader from the roof to the building storm drain or other means of disposal and is located outside of the building

Downstream: refers to a location in the direction of flow after passing a reference point

Drain: any pipe which carries waste water or waterborne wastes in a building drainage system

Drain Field: the area of a piping system arranged in troughs for the purpose of disposing unwanted liquid waste

Drainage Fitting: a type of fitting used for draining fluid from pipes and makes a smooth and continuous interior surface for the piping system

Drainage System: the drainage piping within public or private premises which

conveys sewage, rain water or other liquid wastes to an approved point of disposal, but does not include the mains of a public sewer system or a private or public sewage treatment or disposal plant

Droop: the amount by which the controlled variable pressure, temperature, liquid level or differential pressure deviates from a set value

Drop: refers to piping running to a lower elevation within the same floor level

Drop Elbow: a small elbow having wings cast on each side; the wings have countersunk holes to secure to a ceiling, wall or framing timbers

Drop Tee: a tee having the wings of the same type as the drop elbow

Dross: solid scum that forms on the surface of a metal when molten or melting as a result of oxidation but sometimes because of rising dirt and impurities to the surface

Dry-Pipe Valve: a valve used with a dry-pipe sprinkler system where water is on one side of the valve and air is on the other side; when the fusible link of a sprinkler head melts releasing air from the system, this valve opens, allowing water to flow to the sprinkler head

Dry Weather Flow: sewage collected during the summer which contains little or no ground water by infiltration and no storm water

Durham System: a term used to describe soil or waste systems where all piping is of threaded pipe, tubing or other such rigid construction using recessed drainage fittings to correspond to the type of piping

Durion: a high silicon alloy that is resistant to most corrosive wastes

DWV: type of copper or plastic tubing used for drain, waste or venting pipe

E

Eccentric Fittings: fittings whose openings are offset allowing liquid to flow freely

Effective Openings: the minimum cross-sectional area at the point of water-supply discharge

Effluent: sewage, treated or partially treated, flowing out of sewage treatment equipment

Elastic Limit: the greatest stress which a material can withstand without a permanent deformation after release of the stress

Elbow: a fitting that makes a 90° angle between adjacent pipes unless another angle is specified

Electrolysis: the process of producing chemical changes by passage of an electric current through an electrolyte

End Connection: the method of connecting the parts of a piping system

Engineered Plumbing System: plumbing system designed by using scientific engineering design criteria other than normally given in plumbing codes

Erosion: the gradual destruction of metal or other material by the abrasive action of liquids, gases, solids, etc.

Evapotranspiration: loss of water from the soil by evaporation and by transpiration from growing plants

Existing Work: a plumbing system or any part thereof which has been installed prior to the effective date of applicable code

Extra Heavy: description of piping material, usually cast iron, indicating thicker than standard

Expansion Joint: a joint whose primary purpose is to absorb longitudinal thermal expansion in the pipe line due to heat

Expansion Loop: a large radius bend in a pipe line to absorb longitudinal expansion in the line due to heat

F

Face To Face Dimensions: the dimensions from the face of the inlet port to the face of the outlet port of a valve or fitting

Female Thread: internal thread in pipe fittings, valves, etc.

Filter: device through which fluid is passed to separate contaminates from it

Filter Element or Media: a porous device which performs the process of filtration

Fire Hydrant Valve: a valve that drains at an underground level to prevent freezing when closed

Fire Pumps:

Can Pump: a vertical shaft turbine-type pump in a suction vessel used to raise water pressure

Centrifugal Pump: a pump in which the pressure is developed by the action of centrifugal force

End Suction Pump: a single suction pump having its

suction nozzle on the opposite side of the casing from the stuffing box and having the face of the suction nozzle perpendicular to the longitudinal axis of the shaft

Excess Pressure Pump: low flow, high head pump for sprinkler systems not being supplied from a fire pump; pump pressurizes sprinkler system so that loss of water supply pressure will not cause a false alarm

Fire Pump: pump with driver, controls and accessories used for fire protection service; fire pumps are centrifugal or turbine type with electric motor or diesel engine driver

Horizontal Pump: the shaft normally in a horizontal position

Horizontal Split-Case Pump: a centrifugal pump characterized by a housing which is split parallel to the shaft

In-Line Pump: a centrifugal pump whose drive unit is supported by the pump having its suction and discharge flanges on approximately the same center line

Pressure Maintenance (Jockey) Pump: a pump with

controls and accessories used to maintain pressure in a fire protection system without the operation of the fire pump

Vertical Shaft Turbine Pump: a centrifugal pump with one or more impellers discharging into one or more bowls and a vertical educator or column pipe used to connect the bowl(s) to the discharge head on which the pump driver is mounted

Fitting: the connector or closure for fluid lines

Fitting, Compression: a fitting designed to join pipe or tubing by means of pressure or friction

Fitting, Flange: a fitting which utilizes a radially extending collar for sealing and connection

Fitting, Welded: a fitting attached by welding

Fixture Branch: a pipe connecting several fixtures

Fixture Carrier: a metal unit designed to support a plumbing fixture off the floor

Fixture Carrier Fittings: special fittings for wall mounted fixture carriers

Fixture Drain: the drain from the trap of a fixture to the junction of that drain with any other drain pipe

Fixture Supply: a water-supply pipe connecting the fixture with the fixture branch or directly to a main water supply pipe

Fixture Unit: a measure of probable discharge into the drainage system by various types of plumbing fixtures

Fixture Unit Flow: a measure of the probable hydraulic demand on the water supply by various types of plumbing fixtures

Flange: a ring-shaped plate on the end of a pipe at right angles to the end of the pipe and provided with holes for bolts to allow fastening of the pipe to a similarly equipped adjoining pipe

Flange Bonnet: a valve bonnet having a flange through which bolts connect it to a matching flange on the valve body

Flange Ends: a valve or fitting having flanges for joining to other piping elements

Flange Faces: pipe flanges which have the entire surface of the flange faced straight across, using a full face or ring gasket

Flap Valve: a non-return valve in the form of a hinged disk or flap, sometimes having leather or rubber faces

Flash Point: the temperature at which a fluid gives off sufficient flammable vapor to ignite

Float Valve: a valve which is operated by means of a bulb or ball floating on the surface of a liquid within a tank

Flood Level Rim: the top edge of the receptacle from which water overflows

Flooded: a condition when the liquid rises to the flood-level rim of the fixture

Flow Pressure: the pressure in the water supply pipe near the water outlet while the faucet or outlet is fully open

Flue: an enclosed passage, normally vertical, for removal of gaseous products of combustion to the outer air

Flushing Type Floor Drain: a floor drain which is equipped with an integral water supply, enabling flushing of the drain receptor and trap

Flushometer Valve: a device which discharges a predetermined quantity of water to fixtures for flushing purposes and is actuated by direct water pressure

Foot Valve: a check valve installed at the base of a pump suction pipe to maintain pump prime by preventing pumped liquid from draining away

Footing: the part of a foundation wall or column resting on the bearing soil, rock or piling which transmits the superimposed load to the bearing material

French Drain: a drain consisting of an underground passage made by filling a trench with loose stones and covering with earth; also known as rubble drain

Fresh-Air Inlet a vent line connected with the building drain just inside the house trap and extending to the outer air; it provides fresh air to the lowest point of the plumbing system and with the vent stacks provides a ventilated system; a fresh air inlet is not required on a septic-tank system

Frostproof Closet: a hopper that has no water in the bowl and has the trap and the control valve for its water supply installed below the frost line

Fusion Weld: joining metals by fusion, using oxyacetylene or electric arc

GLOSSARY (cont.)

G

Galvanic Action: when two dissimilar metals are immersed in the same electrolytic solution and connected electrically, there is an interchange of atoms carrying an electric charge between them; the anode metal with the higher electrode potential corrodes; the cathode is protected; magnesium will protect iron, and iron will protect copper

Galvanized Pipe: steel pipe coated with zinc to resist corrosion

Galvanizing: a process where the surface of iron or steel piping or plate is covered with a layer of zinc

Gate Valve: a valve employing a gate, often wedge-shaped, allowing fluid to flow when the gate is lifted from the seat; such valves have less resistance to flow than globe valves

Globe Valve: globe-shaped body with a manually raised or lowered disc which when closed rests on a seat to prevent passage of a fluid

Grade: the slope or fall of a line of pipe in reference to a horizontal plane; in drainage it is expressed as the fall in a fraction of an inch or percentage slope per foot length of pipe

Ground Joint: where the parts to be joined are precisely finished and then ground in so that the seal is tight

H

Header: a large pipe or drum into which each of a group of boilers is connected; also used for a large pipe from which a number of smaller ones are connected in line from the side of the large pipe

Horizontal Branch: a drain pipe extending laterally from a soil, waste stack or building drain

Horizontal Pipe: any pipe or fitting which is installed in a horizontal position

Hub and Spigot: piping made with an enlarged diameter or hub at one end and plain or spigot at the other end; the joint is made tight by oakum, lead or by use of a neoprene gasket inserted in the hub around the spigot

Hubless: soil piping with plain ends; the joint is made tight with a stainless steel or cast iron clamp and neoprene gasket assembly

I

Indirect Waste Pipe: a pipe that does not connect directly with the drainage system but discharges into a plumbing fixture or receptacle that is directly connected to the drainage system

Individual Vent: a pipe installed to vent a fixture trap that connects with the vent system above the fixture served or terminates in the open air.

Induced Siphonage: loss of liquid from a fixture trap due to pressure differential between inlet and outlet of trap, often caused by discharge of another fixture

Industrial Waste: all liquid or waterborne waste from industrial or commercial processes

Insanitary: a condition which is contrary to sanitary principles or is unhealthy

Interceptor: a device that separates and retains hazardous or undesirable matter from normal wastes and permits normal liquid wastes to discharge into the disposal terminal by gravity

Invert: the lowest point on the interior of a horizontal pipe

L

Labeled: equipment or materials bearing a label of a listing agency

Lapped Joint: a pipe joint made by using loose flanges on lengths of pipe whose ends are lapped over to produce a bearing surface for a gasket or metal-to-metal joint

Lap Weld Pipe: made by welding along a scarfed longitudinal seam in which one part is overlapped by the other

Lateral Sewer: a sewer which does not receive sewage from any other common sewer except house connections

Leaching Well: a pit or receptacle having porous walls which permit the contents to seep into the ground; a.k.a. dry well

Leader: the water conductor from the roof to the building storm drain; a.k.a. downspout

Lead Joint: a joint made by pouring molten lead into the space between a bell and spigot and making the lead tight by caulking

Lip Union: a union characterized by a lip that prevents the gasket from being squeezed into the pipe to obstruct flow

Liquid Waste: the discharge from any fixture, appliance or appurtenance in connection with a plumbing system which does not receive fecal matter

Listed: equipment or materials included in a list published by an organization that maintains periodic inspection on current production; the listing states that the equipment or material complies with approved standards or has been tested and found suitable for use in a specified manner

Listing Agency: an agency accepted by the administrative authority which lists and maintains a periodic inspection program on current production

Load Factor: the percentage of the total connected fixture unit flow which is likely to occur at any point in the drainage system

M

Main: the principal artery of the system of continuous piping to which branches may be connected

Main Vent: a vent header to which vent stacks are connected

Malleable: capable of being extended or shaped by beating with a hammer or by rolling pressure

Malleable Iron: cast iron that is heat-treated to reduce brittleness allowing the material to stretch slightly

Manifold: a fitting with a number of branches in line connecting to smaller pipes; an interchangeable term with header

Master Plumber: an individual who is licensed and authorized to install and assume responsibility for contractual agreements pertaining to plumbing and to secure any required permits

Medium Pressure: means valves and fittings are suitable for a working pressure of 125 to 175 psi

Mill Length: also known as random length; run-of mill pipe is 16 to 20 ft in length; some pipe is made in double lengths of 30 to 35 ft

N

Needle Valve: a valve provided with a long tapering point in place of the ordinary valve disk; the tapering point permits fine graduation of the opening

Nipple: a tubular pipe fitting normally threaded on both ends and is under 12 inches in length; pipe over 12 inches is regarded as cut pipe

GLOSSARY (cont.)

O

O.D. Pipe: is pipe that measures over 14 inches N.P.S., where the nominal size is the outside diameter and not the inside diameter.

Offset: a combination of pipe and/or fittings which join two nearly parallel sections of a pipe line

Outfall Sewers: sewers receiving the sewage from the collection system and carrying it to the point of final discharge or treatment

Oxidized Sewage: sewage in which the organic matter has been combined with oxygen resulting in natural stability

P

Percolation: the seeping of a liquid downward through a filtering medium which may or may not fill the pores of the medium

Pitch: the amount of slope or grade given to horizontal piping and expressed in inches of vertically projected drop per foot

Plug Valve: one with a short section of a cone or tapered plug through which a hole is cut so that fluid can flow through when the hole lines up with the inlet and outlet; when the plug is rotated, flow is blocked

Plumbing: the practice, materials and fixtures used in the installation, maintenance and alteration of all piping, fixtures, appliances and appurtenances in connection with the following: sanitary or storm drainage facilities; the venting system; the public or private water-supply systems within or adjacent to any building or structure; water supply systems; the storm water liquid waste or sewage system of any premises to their connection with any point of public disposal

Plumbing Appliance: a plumbing fixture which is intended to perform a special function; its operation may be dependent upon one or more energized components such as motors, controls, heating elements or pressure/temperature-sensing elements; these fixtures may operate automatically through one or more of the following actions: a time cycle, temperature range, pressure range, a measured volume or weight; or the fixture may be manually controlled by the user

Plumbing Appurtenances: a manufactured device, prefabricated assembly or job assembled component parts which are added to the basic plumbing system; an appurtenance demands no additional water supply nor does it add any discharge load to a fixture or the drainage system; it performs some useful function in the operation, maintenance or safety of the plumbing system

Plumbing Engineering: the application of scientific principles to the design, installation and operation of distribution systems for the transport of liquids and gases

Plumbing Fixtures: installed receptacles, devices or appliances which are supplied with water or which receive liquid or liquid borne wastes and discharge such wastes into the drainage system to which they may be directly or indirectly connected; industrial or commercial tanks, vats, etc. are not plumbing fixtures, but may be connected to or be discharged into approved plumbing fixtures such as traps

Plumbing Inspector: any person who, under the supervision of the department having jurisdiction, is authorized to inspect plumbing and drainage as defined in the code for the municipality

Plumbing System: all potable water supply and distribution pipes, plumbing fixtures and traps, drainage and vent pipe, and all building (house) drains including their respective joints and connections, devices, receptacles and appurtenances within the property lines of the premises; these include potable water piping, potable water treating or using equipment, gas piping, water heaters, etc.

Polymer: a chemical compound or mixture of compounds formed by polymerization and consisting essentially of repeating structural units

Pool: a water receptacle used for swimming or bathing designed to accommodate more than one person at a time

Potable Water: water which is satisfactory for drinking, cooking and domestic purposes

Precipitation: the total measurable supply of water received directly from clouds as snow, rain, hail and sleet expressed in inches per day, month or year

Private Sewage Disposal System: a septic tank with the effluent discharging into a subsurface disposal field, one or more seepage pits or a combination of both

Private Sewer: a sewer which is privately owned and not directly controlled by public authority

Private Use: plumbing fixtures in residences and apartments, private bathrooms in hotels and hospitals, rest rooms in commercial establishments containing restricted-use single fixture or groups of single fixtures and similar installations where the fixtures are intended for the use of a family or an individual

Public Sewer: a common sewer directly controlled by public authority

Public Use: applies to locked and unlocked toilet rooms and bathrooms used by employees, occupants, visitors or patrons in or about any premises

Putrefacation: biological decomposition of organic matter with the production of foul-smelling products and usually happens when there is a deficiency of oxygen

R

Receptor: a plumbing fixture or device of such material, shape and capacity that adequately receives the discharge from indirect waste pipes that are made and located to be cleaned easily

Reduced Size Vent: dry vents which are smaller than those allowed by plumbing codes

Reducer: a pipe fitting with inside threads that are larger at one end than at the other

Relief Valve: designed to open automatically to relieve excess pressure

Relief Vent: a vent designed to provide circulation of air between drainage and vent systems or to act as an auxiliary vent

Residual Pressure: pressure remaining in a system while water is being discharged from outlets

Resistance Weld Pipe: pipe made by bending a plate into circular form and passing electric current through the material to obtain a welding heat

Return Offset: a double offset installed to return the pipe to its original alignment

Revent Pipe: that part of a vent pipe line which connects directly with an individual waste or group of wastes, underneath or back of the fixture, and extends either to the main or branch vent pipe

Rim: an unobstructed open edge of a fixture

Riser: a water supply pipe which extends vertically one full story or more to convey water to branches or fixtures; a vertical pipe used to carry water for fire protection to elevations above or below grade

Rolling Offset: same as offset, but used where the two lines are not in the same vertical or horizontal plane

Roof Drain: a drain installed to remove water collecting on the surface of a roof and to discharge it into the leader

Rotary Pressure Joint: a joint for connecting a pipe under pressure to a rotating machine

Roughing-In: the installation of all parts of the plumbing system which can be completed prior to the installation of fixtures; this includes drainage, water supply and vent piping, and the necessary fixture supports

Run: a length of pipe made up of more than one piece; a portion of a fitting having its ends in line, in contradistinction to the branch or side opening, as of a tee

S

Saddle Flange: a flange curved to fit a boiler or tank and to be attached to a threaded pipe; the flange is riveted or welded to the boiler or tank

Sand Filter: a water treatment device for removing solid or colloidal material with sand as the filter media

Sanitary Sewer: the conduit or pipe carrying sanitary sewage; it may include storm water and the infiltration of ground water

Saturated Steam: steam at the same temperature as water boils under the same pressure

Screwed Flange: a flange screwed on a pipe which it connects to adjoining pipe

Screwed Joint: a pipe joint consisting of threaded male and female parts screwed together

Seamless Pipe: pipe or tube formed by piercing a billet of steel and then rolling

GLOSSARY *(cont.)*

Seepage Pit a lined excavation in the ground which receives the discharge of a septic tank designed to permit the effluent from the septic tank to seep through its bottom and sides

Septic Tank: a watertight receptacle which receives the discharge of a drainage system so as to separate solids from the liquid and digest organic matter through a period of detention

Service Fitting: a street ell or street tee with male threads at one end and female threads at the other

Service Pipe: a pipe connecting water or gas mains with a building

Set: same as offset, but used in place of offset where the connected pipes are not in the same vertical or horizontal plane

Setback: in a pipe bend, the distance measured back from the intersection of the centerlines to the beginning of the bend

Sewage: any liquid waste containing animal, vegetable or chemical wastes in suspension or solution

Sewage Ejector: a mechanical device or pump for lifting sewage

Short Nipple: one whose length is a little greater than that of two threaded lengths or somewhat longer than a close nipple so that it has some unthreaded portion between the two threads

Shoulder Nipple: a nipple of any length which has a portion of pipe between two threads; it is halfway between the length of a close nipple and a short nipple

Siamese: a hose fitting for combining the flow from two or more lines into a single stream

Side Vent: a vent connected to the drain pipe through a fitting at an angle not greater than 45 degrees to the vertical

Sleeve Weld: a joint made by butting two pipes together and welding a sleeve over the outside

Slip-On Flange: a flange slipped over the end of the pipe and then welded to the pipe

Sludge: the accumulated suspended solids of sewage deposited in tanks, beds or basins and is mixed with water to form a semi-liquid

Socket Weld: a joint made by use of a socket weld fitting which has a prepared female end or socket for insertion of the pipe to which it is welded

GLOSSARY *(cont.)*

Soil Pipe: any pipe which conveys the discharge of water closets, urinals or fixtures, with or without the discharge from other fixtures, to the building drain or sewer

Solder Joint: a method of joining tube by use of solder

Special Wastes: wastes which require some special method of handling such as the use of indirect waste piping and receptors, corrosion resistant piping, sand, oil or grease interceptors, condensers or other pre-treatment facilities

Spiral Pipe: pipe made by coiling a plate into a helix and riveting or welding the overlapped edges

Sprinkler System: an integrated system of underground and overhead piping designed in accordance with fire protection engineering standards

Sprinkler System Classification:

1. wet-pipe systems
2. dry-pipe systems
3. pre-action systems
4. deluge systems
5. combined dry-pipe and pre-action systems

Stack: the vertical main of a system of soil, waste or vent piping extending through one or more stories

Stack Group: the location of fixtures in relation to the stack so that by means of proper fittings, vents may be reduced to a minimum

Stack Vent: the extension of a soil or waste stack above the highest horizontal drain connected to the stack; also known as waste or soil vent

Stack Venting: a method of venting a fixture or fixtures through the soil or waste stack

Stainless Steel Pipe: an alloy steel pipe with corrosion-resisting properties, usually imparted by nickel and chromium

Standard Pressure: formerly used to designate cast-iron flanges, fittings, valves, etc., suitable for a maximum working steam pressure of 125 psi

Standpipe: a vertical pipe generally used for the storage and distribution of water for fire extinguishing purposes

Standpipe System: an arrangement of piping, valves, hose connections and equipment installed in a structure with the hose connections located in such a manner that water can be discharged in streams or spray patterns through attached hose and nozzles for the purpose of extinguishing a fire; this is accomplished by connections to water supply systems or by pumps, tanks and other equipment necessary to provide an adequate supply of water to the hose connections

Storm Sewer: a sewer used for conveying rain water, surface water condensate, cooling water or similar liquid wastes exclusive of sewage and industrial waste

Stop Valve: a valve used for the control of water supply to a single fixture

Strain: change of shape or size of body produced by stress

Street Elbow: an elbow with male thread on one end, and female thread on the other

Stress: reactions within the body resisting external forces acting on it

Subsoil Drain: a drain which receives only subsurface or seepage water and conveys it to a place of disposal

Sub-Main Sewer: a sewer into which the sewage from two or more lateral sewers is discharged; also known as a branch sewer

Sump: a tank or pit which receives sewage or liquid waste located below the normal grade of the gravity system and which must be emptied by mechanical means.

Sump Pump: a mechanical device for removing liquid waste from a sump

Superheated Steam: steam at a higher temperature than that at which water would boil under the same pressure

Supervisory Switch: a device attached to a valve, which, when the valve is closed, will annunciate a trouble signal at a remote location

Supports: devices for supporting and securing pipe and fixtures to walls, ceilings, floors or structural members

Swimming Pool: a structure, basin or tank containing water for recreation

Swing Joint: an arrangement of screwed fittings and pipe that provides for expansion

Swivel Joint: one employing a special fitting that is pressure tight under movement of the device to which it is connected

T

Tee: a fitting, either cast or wrought, that has one side outlet at right angles to the run

Tempered Water: water ranging in temperature from 85°F (29°C) up to 110°F (43°C)

Trailer Park Sewer: the horizontal piping of a drainage system which begins two feet downstream from the last trailer site connection, receives the discharge of the trailer site, and conveys it to a public sewer, private sewer, individual sewage disposal system or other point of disposal

Trap: a fitting designed to provide, when properly vented, a liquid seal which will prevent the back passage or air without significantly affecting the flow of sewage or waste water through it

Trap Primer: a device or system of piping to maintain a water seal in a trap

Trap Seal: the maximum vertical depth of liquid that a trap will retain

Travel: see offset

Turbulence: any deviation from parallel flow in a pipe due to rough inner wall surfaces, obstructions or directional changes

U

Underground Piping: piping in contact with the earth below grade

Union: a device used to connect pipes and usually consisting of three pieces: the thread end fitted with exterior and interior threads; the bottom end fitted with interior threads and a small exterior shoulder; and the ring which has an inside flange at one end while the other end has an inside thread like that on the exterior of the thread end; unions permit connections and disconnections with little disturbance to pipe sections

Union Ell: an ell with a male or female union at one end

Union Joint: a pipe coupling, usually threaded, which permits disconnection without disturbing other sections

Union Tee: a tee with a male or female union at one end of the run

Upstream: term referring to a location in the direction of flow before reaching a reference point

V

Vacuum: any pressure less than that exerted by the atmosphere and may be termed a negative pressure

Vacuum Breaker: a backflow preventer

Vacuum Relief Valve: a device to prevent excessive vacuum in a pressure vessel

Vent, Loop: any vent connecting a horizontal branch or fixture drain with the stack vent of the originating waste or soil stack

Vent Stack: a vertical vent pipe installed primarily for the purpose of providing circulation of air to and from any part of the drainage system

Vertical Pipe: any pipe or fitting which is installed in a vertical position or which makes an angle of not more than 45 degrees with the vertical

Vitrified Sewer Pipe: conduit made of fired and glazed earthenware installed to receive sewage or waste

W

Waste: the discharge from any fixture, appliance or appurtenance which does not contain fecal matter

Waste Pipe: the discharge pipe from any fixture, appliance or appurtenance in connection with the plumbing system which does not contain fecal matter

Water Conditioner: a device which conditions or treats a water supply to change its chemical content or remove suspended solids by filtration

Water-Distributing Pipe: a pipe which conveys potable water from the building supply pipe to the plumbing fixtures and water outlets

Water Hammer: the noise and vibration which develop in a piping system when a column of non-compressible liquid flowing through a pipe line at a given pressure and velocity is abruptly stopped

Water Hammer Arrester: a device, other than an air chamber, designed to provide protection against excessive surge pressure (hammering)

Water Main: the water supply pipe for public or community use; normally under the jurisdiction of the municipality or water company

Water Riser: a water supply pipe which extends vertically one full story or more to convey water to branches or fixtures

GLOSSARY *(cont.)*

Water-Service Pipe: the pipe from the water main or other source of water supply to the building served

Water Supply System: the building supply pipe, the water-distributing pipes and the necessary connecting pipes, fittings, control valves and all appurtenances carrying or supplying potable water in or adjacent to the building or premises

Welding-End Valves: valves without end flanges and with ends tapered and beveled for butt welding

Welding Fittings: wrought or forged-steel elbows, tees, reducers, etc., beveled for welding to pipe

Welding Neck Flange: a flange with a long neck beveled for butt welding to pipe

Wet Vent: a vent which also serves as a drain

Wiped Joint: a lead pipe joint in which molten solder is poured after scraping and fitting the parts together; the joint is wiped up by hand with a moleskin or cloth pad while the metal is in a plastic condition

Wrought Iron: iron refined to a plastic state in a puddling furnace; it is characterized by the presence of 3 percent slag irregularly mixed with pure iron and 0.5 per cent carbon

Wrought Pipe: refers to both wrought steel and wrought iron; wrought means worked, as in the process of forming furnace-welded pipe from skelp, or seamless pipe from plates or billets; wrought pipe is thus used as a distinction from cast pipe

Wye (Y): a fitting, either cast or wrought, that has one side outlet at any angle other than 90 degrees

Y

Yoke Vent: a pipe connecting upward from a soil or waste stack to a vent stack for the purpose of preventing pressure changes in the stacks

About The Author

Paul Rosenberg has an extensive background in the construction, data, electrical, HVAC and plumbing trades. He is a leading voice in the electrical industry with years of experience from an apprentice to a project manager. Paul has written for all of the leading electrical and low voltage industry magazines and has authored more than 30 books.

In addition, he wrote the first standard for the installation of optical cables (ANSI-NEIS-301) and was awarded a patent for a power transmission module. Paul currently serves as contributing editor for *Power Outlet Magazine*, teaches for Iowa State University and works as a consultant and expert witness in legal cases. He speaks occasionally at industry events.